Private Pilot Test

Biblioteca Aeronáutica
aviación en simples pasos

Conforti, Facundo
 / Facundo Conforti. - 1a ed. - Mar del Plata : Facundo Jorge Conforti, 2023.
 194 p. ; 21 x 15 cm.

 1. Aviación. I. TÌtulo.
 CDD 629.1343

1ra Edición.

Biblioteca Aeronáutica
aviación en simples pasos

Facundo Conforti, 2024.
All rights reserved

Published in Argentina. This handbook is sold subject to the condition that it shall not, by way of trade or otherwise, be lent, re-sold, hired out, or otherwise circulated without the publisher's prior consent.

Introduction

If you are preparing to take the Private Pilot License (PPL) exam, this book will be of significant assistance. Throughout its pages, we will test your knowledge necessary to obtain your first aviation license, your Private Pilot License (PPL).

The format of this book is divided into two sections. The first section is based on a question bank related to the minimum theoretical knowledge a student should have to pass their exam. The second section features an innovative system for simulating the practical flight exam with the examiner (inspector from the aviation authority), where you must carefully select the steps to follow to successfully pass your exam.

This unprecedented work is unique in the market. You will learn everything necessary to pass your actual exam and obtain your first aviation license.

Capt. Facundo Conforti

Biblioteca Aeronáutica
aviación en simples pasos

Chapter 1

Theoretical Knowledge for PPL

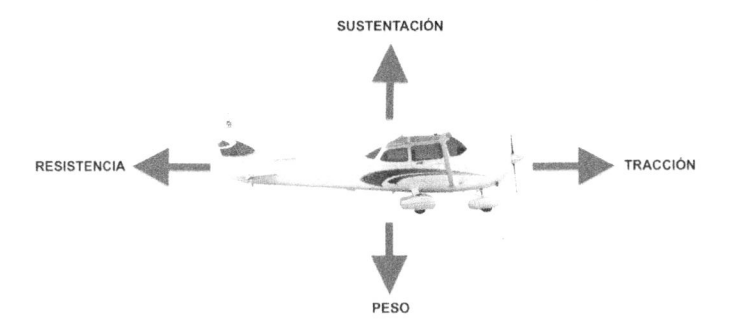

Basic Aeronautical Knowledge

The beginning of this question guide is based on basic aeronautical knowledge related to the aircraft, its different basic maneuvers, and an initial aeronautical environment for those just taking their first steps in the career of a professional pilot.

01.- The control axes of an aircraft are:

    A. Vertical Axis (yaw). Longitudinal Axis (roll). Transverse Axis (pitch).
    B. Vertical Axis (yaw). Longitudinal Axis (roll). Transverse Axis (pitch).
    C. Vertical Axis (pitch). Longitudinal Axis (roll). Transverse Axis (yaw).

The control axes of an aircraft allow for the three main movements for flight maneuvers: yaw movement, based on the vertical axis; roll movement, based on the longitudinal axis; and pitch movement, based on the transverse axis.

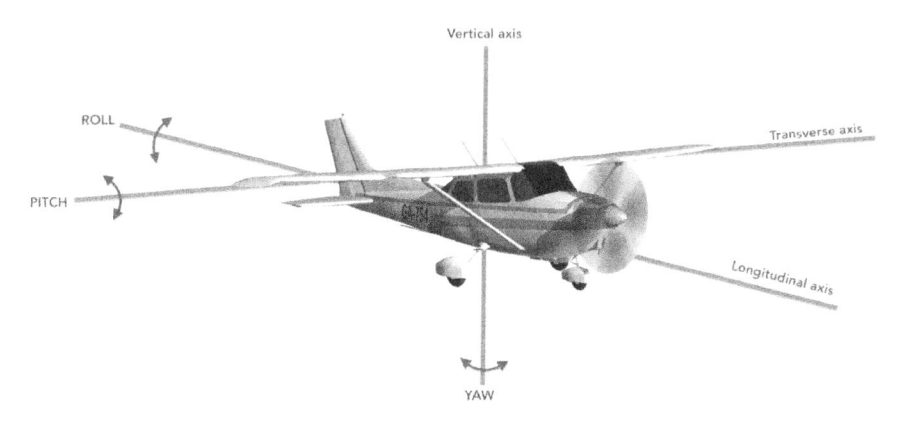

02.- The control surfaces that operate on the axes are:

A. Elevator. Flap. Ailerons. Rudder
B. Rudder. Flap. Ailerons.
C. Elevator. Rudder. Ailerons.

Remember! There are control surfaces known as primary and secondary. The primary surfaces are directly related to the control axes (ailerons, rudder, and elevator). On the other hand, the secondary surfaces are complementary surfaces such as flaps, spoilers, slats, among others.

03.- Control surface located at the tail of the aircraft, parallel to the position of the ailerons. It controls the "pitch" movement.

A. Elevator
B. Rudder
C. Ailerons and Flaps.

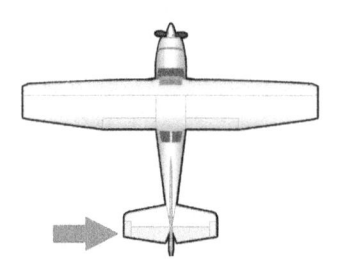

The elevator is located at the tail of the aircraft, parallel to the position of the ailerons. This control surface governs the movement performed on the "transverse axis" of the aircraft, known as "pitch." It is the primary control surface for executing climb and descent procedures. From the cockpit, the elevator is controlled using the control column, moving it forward or backward depending on the desired procedure.

04.- The runway numbering is given by numbers and, in some cases, includes letters (L-left, C-center, R-right), indicating:

    A. Su Its orientation with respect to true north
    B. Its orientation with respect to magnetic north
    C. Its orientation with respect to other runways and the airport

The numbering indicates the magnetic orientation of the runway concerning the degrees of the "Compass Rose." For example, a runway oriented to 230° magnetic would be numbered as such. If it were oriented towards the EAST, its numeric designation would be 09, if towards the south it would be 18, and so on, covering the 360°. Note that only the first two digits of the orientation are used, eliminating the last number.

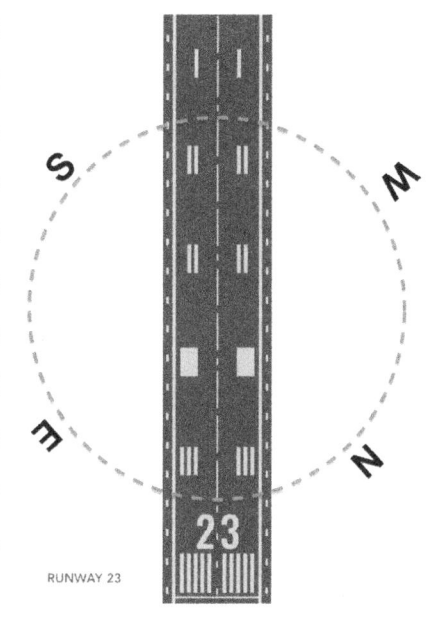

RUNWAY 23

05.- A traffic pattern is composed of:

A. An upwind leg, a base leg, and a final leg
B. The runway, an upwind leg, a base leg, and a final leg
C. An upwind leg, a base leg, a final leg, and a go-around

The traffic pattern starts with an upwind leg, followed by a base leg perpendicular to the wind direction, and ends with a final leg on a track opposite the wind direction.

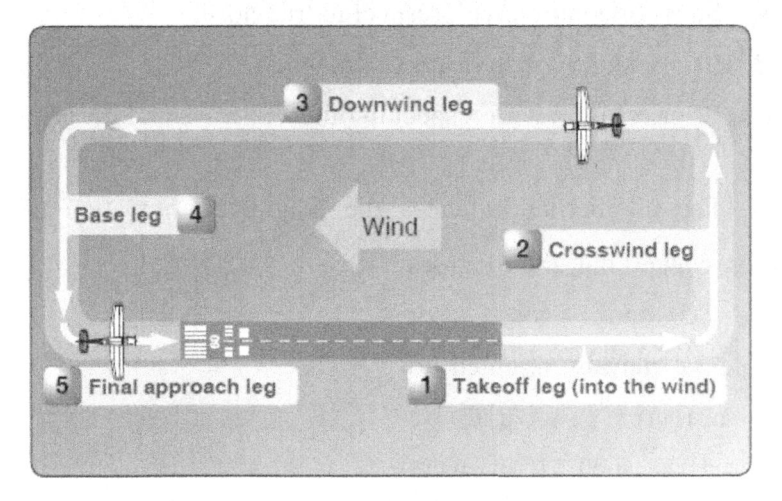

06.- Regarding the traffic pattern, the pilot must:

A. AtLand after the final leg, facing the wind
B. AtLand after the final leg, regardless of the wind
C. Land after the final leg, with a tailwind

Remember! Landings should be performed on a flight path opposite the wind direction to optimize the airflow over the wing profile.

07.- The taxiing of an aircraft is commanded by:

A. The pedals (elevator), brakes, and power.
B. The brakes, ailerons (rudder), and power.
C. The pedals, power, and brakes.

Remember! During taxiing, the pilot controls the direction of the aircraft using the pedals, power control, and brakes. In the case of taxiing with a crosswind, the ailerons may also be used to compensate for the force exerted by the wind.

08.- A takeoff consists of:
A. Aligning on the runway and the takeoff roll.
B. The takeoff roll, rotation, and initial climb.
C. Aligning on the runway, the takeoff roll, and rotation.

Remember! Although aligning on the runway before takeoff is an optional procedure, it is recommended to achieve an orderly takeoff in its three phases: takeoff roll, rotation, and initial climb.l.

09.- Regarding the bank angle of turns, they are defined as:

A. Smooth turn 15°. Normal turn 20°. Steep turn 30°.
B. Smooth turn 10°. Normal turn 20°. Steep turn 40°.
C. Smooth turn 15°. Normal turn 30°. Steep turn 45°.

The bank angle of turns is determined every 15°.

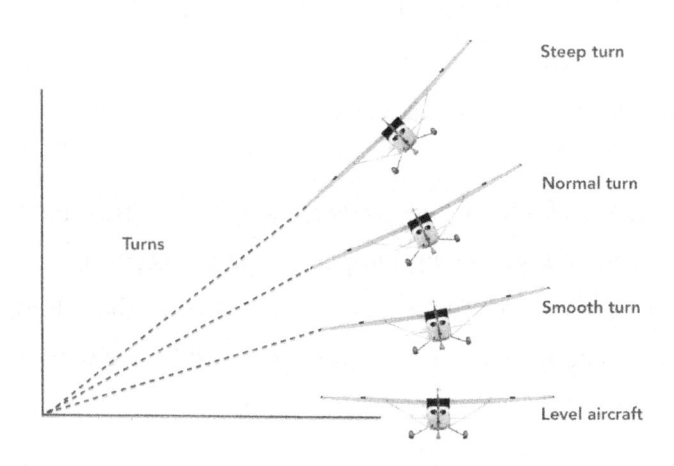

Steep turn

Normal turn

Turns

Smooth turn

Level aircraft

10.- The stall speed is defined as:

    A. The speed at which the aircraft stops generating lift.

    B. The speed at which the wing stops producing lift.

    C. The speed at which the aircraft is approaching a stall.

> **Remember!** The stall speed is the value at which the wings can no longer generate lift due to an excessive angle of attack.

11.- The pilot should practice the stall approach maneuver to:

    A. Identify the speed at which the aircraft stops generating lift.

    B. Identify the abnormal behavior of the aircraft.

    C. Learn the stall recovery procedure.

Remember! It is crucial for the pilot to recognize a stall situation by identifying the abnormal behavior of the aircraft.

12.- The "Stall" audible warning in an aircraft indicates:

    A. That the aircraft is in a stall.
    B. That the aircraft is nearing a stall.
    C. That the aircraft is not generating lift.

Remember! The stall audible warning indicates that the wings are losing lift and alerts the pilot to take immediate action.

13.- The first step in recovering from a stall is:

    A. Increasing speed by adding power.
    B. Reducing the angle of attack to increase speed.
    C. Reducing drag to increase lift.

Remember! The first and most immediate action by the pilot should be to reduce the angle of attack and increase speed as a result.

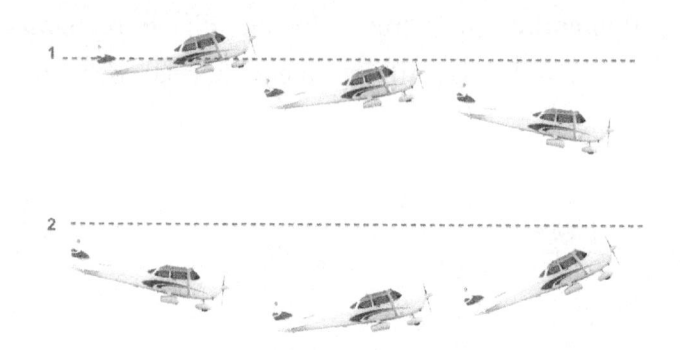

14.- Regarding the approach before landing, there are:
   A. Approaches of 90°, 180°, and 360°.
   B. A single final approach to the active runway.
   C. An approach according to the traffic pattern.

Remember! The aircraft approach must adhere to the traffic pattern. Within this pattern, it is possible to perform an approach of 90°, 180°, and 360°.

15.- The "FLARE" maneuver is defined as:
   A. A change in the angle of attack during the approach.
   B. A change in the aircraft's flight attitude.
   C. A controlled reduction in lift upon landing.

Remember! The Flare is a maneuver where the pilot changes the angle of attack during the final approach, just before making contact with the runway, to achieve a smooth landing on the surface.

16.- Regarding the "FLARE" maneuver, it should be performed:

    A. On the final leg, just before entering the runway.
    B. On the final leg, just above the runway threshold.
    C. Over the runway, after passing the threshold.

> **Remember!** It is crucial to perform the Flare maneuver once the aircraft is over the runway to ensure a safe landing.

17.- Regarding the landing roll during a crosswind operation:a la carrera de aterrizaje durante una operación con viento cruzado:

    A. The pilot should keep the aircraft centered on the runway by using the rudder pedals.
    B. The pilot should keep the aircraft within the runway limits, regardless of the center, due to the effect of the wind.
    C. Both answers are correct but involve using ailerons and directional control.

> **Remember!** The pilot should control the direction of the aircraft by applying pressure on the pedals to ensure that the landing roll remains centered on the runway.

Aerodynamics

Aerodynamics is the science that studies the behavior of a body immersed in and moving through a mass of air. Therefore, it is crucial for the pilot to understand the effects of this body or "WING" when exposed to a free air stream.

18.- When an object enters an air stream, two aerodynamic forces are generated:

- A. Weight and thrust.
- B. Lift and loss of lift.
- C. Lift and drag.

Remember! When the wing surface enters an air stream, lift and drag are produced.

19- An air particle has two variables, which are:

- A. Pressure and weight.
- B. Drag and speed.
- C. Speed and pressure.

Every air particle is characterized by pressure and speed. This principle is better understood through Bernoulli's Theorem.

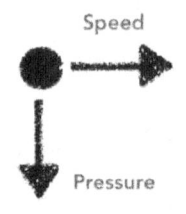

20.- Bernoulli's Theorem states that an air particle

    A. Can increase one of its variables but simultaneously decrease the other to maintain a constant value.

    B. Can increase or decrease both variables to maintain a constant value.

    C. Can alter its variables independently without maintaining a constant value.

Remember! The theorem states that an air particle can change one of its variables, directly affecting the other to maintain a constant value.

21.- The Venturi effect defines that particles passing through a constriction:

    A. Increase their pressure and decrease their drag.

    B. Decrease their drag and increase their speed.

    C. Increase their speed and decrease their pressure.

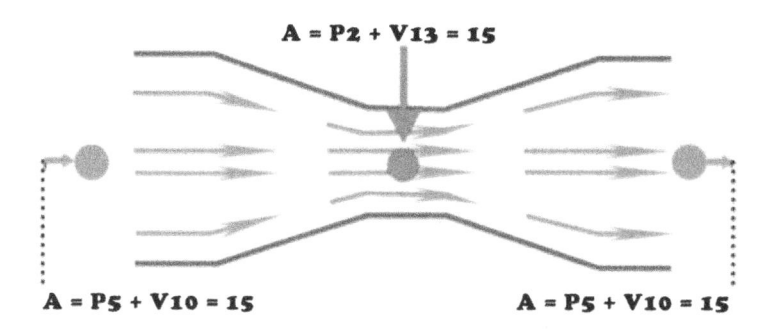

$$A = P_2 + V_{13} = 15$$

$$A = P_5 + V_{10} = 15 \qquad A = P_5 + V_{10} = 15$$

22.- There are 4 forces acting on an aircraft, which are:
  A. Weight, thrust, speed, and drag.
  B. Weight, lift, drag, and thrust.
  C. Weight, speed, lift, and drag.

23.- The angle of attack of a wing is formed by:
  A. The wing cord and the relative wind.
  B. The wing's base or lower surface and the wing cord.
  C. The lower surface and the upper surface of the wing.

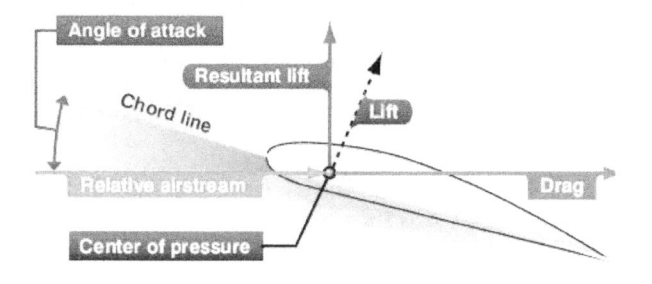

24.- What is the relationship of the 4 forces acting on an aircraft during straight and level flight?
  A. All four forces are balanced.
  B. Lift = Weight, and Thrust = Drag.
  C. Lift = Drag, and Weight = Thrust.

18

Remember! The four forces are balanced because they exert the same pressure to maintain the aircraft in a steady flight attitude, meaning no climb or descent, and at a constant speed, without acceleration or deceleration.

25.- High-lift devices allow:
- A. Increasing lift without increasing speed.
- B. Maintaining lift at a constant speed.
- C. Reducing drag and increasing lift.

Remember! High-lift devices, such as flaps, increase the wing's surface area, resulting in increased lift.

Navigation

We define "Navigation" as the set of techniques that allow the pilot to fly an aircraft from a starting point to a destination, having calculated all the intervening variables.

26.- The difference between a COURSE and a HEADING is given by:

- A. The type of navigation being performed.
- B. The effect of the wind on the mentioned values.
- C. There is no difference, as both are directions.

Navegación con viento cruzado y corrección de rumbo

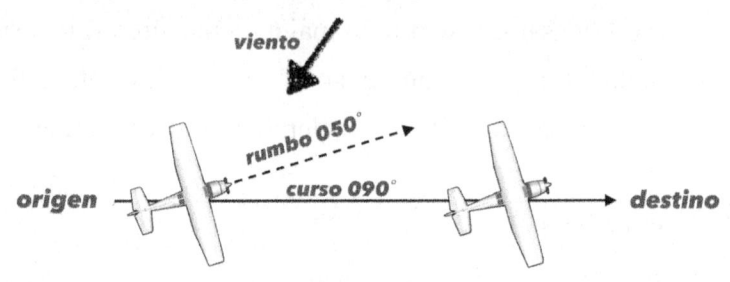

origen — rumbo 050° / curso 090° / viento — destino

27.- Flying at a speed of 90 KT (knots), how long would it take the aircraft to cover a distance of 334 KM?

    A. At a constant speed, 120 minutes.
    B. At a constant speed, 90 minutes.
    C. It depends on the speed in KM, due to the distance in KM.

Remember! In aviation, it is common to use speed in knots (KT) and distance in nautical miles (NM). To convert units, 1 NM is equal to 1.852 KM. Therefore, 334 KM is approximately 180 NM. At a speed of 90 KT, the time required to cover 180 NM is 120 minutes.

28.- Flying at a speed of 100 MPH, how many KM will an aircraft cover in 90 minutes?

    A. 161 KM.
    B. 321 KM.
    C. 241 KM.

Remember! One mile per hour (MPH) is equal to 1.609 KM.

29.- According to the runway on this chart, what could be the orientation of the runway at Tucumán Airport?

   A. Runway 15-33
   B. Runway 25-07
   C. Runway 02-20.

Remember! The graphical representation of runways on a navigation chart corresponds to magnetic north.

30.- What is the correct cruising altitude for flying under visual flight rules with a course of 130°?

   A. Thousands + 500, in EVEN numbers.
   B. Thousands + 500, in ODD numbers.
   C. Thousands, in EVEN or ODD numbers

Remember! Avoid confusion by using the rule "even to the west, and odd to the east," adding 500 feet if in uncontrolled visual flight.

31.- Given the following data: Distance 370 KM, Speed 75 KT, Consumption 35 LT/HR. The results for flight time and fuel consumption are:

   A. Time: 01:50 hrs, Consumption: 84 LT.
   B. Time: 02:40 hrs, Consumption: 94 LT.
   C. Time: 95 minutes, Consumption: 52 LT.

32.- The minimum and legal total fuel for a visual navigation is given by the sum of the following variables:

A. The amount needed to reach the destination + alternate.
B. The amount needed to cover the destination + 45 minutes of reserve.
C. The amount needed to reach the destination + alternate + 45 minutes.

Remember! For safety reasons, every flight must plan for an alternate aerodrome and account for the fuel required to reach it.

Weather

We define "Meteorology" as the science that studies atmospheric processes and all their phenomena. These phenomena directly affect aviation activities, and it is crucial for pilots to have basic knowledge of this science.

33.- The wind at an altitude of 6000 FT comes from the "Northwest," but at the surface, the wind comes from the "North." What is the reason for this variation?

A. Different pressures at different altitudes.
B. The "Coriolis Effect" on the surface.
C. Friction between the wind and the surface.

Remember! At high altitudes, the Earth's surface has minimal influence on the wind. However, in lower atmospheric layers, wind speeds are affected by friction with the surface.

34.- The Dew Point is defined as:

    A. The temperature at which water vapor contained in the air condenses.
    B. The temperature at which an air particle turns into a water droplet.
    C. The temperature at which the atmosphere must cool to precipitate.

Remember! The dew point is the temperature to which air must be cooled for the existing water vapor to begin condensing.

35.- The amount of water vapor that air can hold...
    A. Depends on the dew point.
    B. Depends on the air temperature.
    C. Depends on the air stability.

36.- Clouds and fog are the result of:
    A. Condensed water vapor.
    B. Relative humidity close to 100%.
    C. Presence of water vapor in the air

37.- The cloud family is divided into 4 groups according to:
   A. Their shape and hazard.
   B. Their altitude range.
   C. Their composition.

> **Remember!** Considering that the troposphere is divided into three layers at different heights, clouds can be classified according to the layer they occupy: low, middle, and high clouds.

38.- Which type of cloud presents the most turbulence?
   A. Cumulus Towers.
   B. Cumulonimbus.
   C. Nimbostratus.

> **Remember!** The vertical extent of a cumulus cloud is a good indication of the intensity of vertical airflow inside and beneath the cloud, and consequently, the turbulence within and around the cloud.

39.- How does icing affect lift?
   A. It does not affect lift.
   B. It adheres to the airfoil, changing its shape and affecting its lift capability.
   C. It adheres to the aircraft, increasing its weight and generating more drag, decreasing lift.

Remember! Ice formation on the airfoil modifies its shape and impacts its ability to generate lift.

Flight Instruments

Continuing with the basic concepts of each subject, it's time to cover flight instruments. Flight instruments are classified into two main groups: instruments based on pressure measurement and instruments based on gyroscopic properties.

40.- The instruments based on pressure measurement are:
   A. Altimeter, airspeed indicator, heading indicator, and vertical speed indicator.
   B. Vertical speed indicator, altimeter, airspeed indicator.
   C. Artificial horizon, altimeter, and airspeed indicator

Remember! Instruments based on pressure measurement use the aircraft's static ports and pitot tube.

41.- The green arc shown on an airspeed indicator indicates:
   A. Normal operating range.
   B. Normal operating range with landing gear down.
   C. Normal operating range with flaps extended.

Remember! The green arc around the speed values on an airspeed indicator denotes the aircraft's normal operating range.

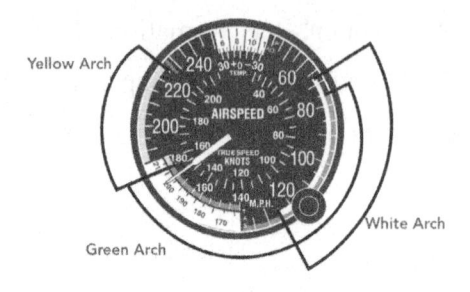

42.- The yellow arc on the airspeed indicator indicates:
   A. Caution range. Not operable.
   B. Caution range. Operable.
   C. Caution range. Operable with landing gear retracted.

43.- The operating principle of an altimeter is based on:
   A. Measuring changes in pressure due to altitude.
   B. Variations in the speed of pressure change.
   C. Variations in pressure during climb and descent.

44.- If the static pressure port is blocked, you lose:
   A. The information from the altimeter, variometer, and directional gyro.
   B. The information from the airspeed indicator and variometer only.
   C. The information from the airspeed indicator, altimeter, and variometer.

   Remember! If a pressure port is blocked, all instruments relying on that information will cease to function or will function abnormally.

45.- Gyroscopic instruments are:
   A. Variometer, artificial horizon, and directional gyro.
   B. Turn coordinator and artificial horizon only.
   C. Directional gyro, artificial horizon, and turn coordinator.

   Remember! The gyroscope is a mechanical device capable of measuring, maintaining, or changing the orientation in space of a mobile object.

46.- Observing the following figure, number 01 indicates:
   A. Visual reference marks for bank angle in a turn.
   B. Visual reference marks for drift correction.
   C. Reference marks for wing leveling in straight flight.

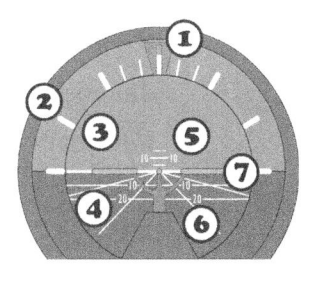

47.- Observing the following figure, numbers 05 and 06 indicate:
  A. Numeric scale for the angle of attack.
  B. Numeric scale for pitch up/down indication.
  C. Numeric scale for climb and descent indication.

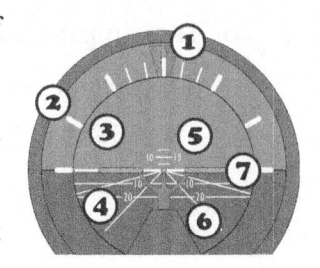

48.- Gyroscopic instruments are based on:
  A. Gyroscopes and the aircraft's electrical system.
  B. The properties of "Rigidity" and "Precession".
  C. Internal and individual properties of each instrument.

49.- For the "Directional Gyro" information to be accurate during a flight, the pilot should:

  A. Calibrate the "Gyro" with the runway heading before takeoff to match the runway axis.
  B. Calibrate the "Gyro" with the current heading when starting the aircraft and before taxiing to avoid errors.
  C. Calibrate the "Gyro" with the compass regularly in flight, as the Gyro can precess.

Remember! It is always advisable to confirm the directional gyro information with the magnetic compass information, to calibrate the gyroscopic instrument for more accurate information.

50.- The altimeter will indicate::

　A. True altitude.
　B. Absolute altitude.
　C. Both, as desired.

51.- Which instruments would become inoperative if the Pitot tube gets blocked?

　A. The altimeter and the airspeed indicator.
　B. Only the altimeter.
　C. Only the airspeed indicator.

Power plant

The powerplant group in an aircraft is responsible for generating the thrust necessary for self-propulsion. It works in conjunction with the propeller's traction system. Basic knowledge of this subject is very important for a pilot, as it will help them understand what is happening inside the engine.

52.- The operating principle of an aircraft engine is based on the well-known "Otto Cycle" or 4-stroke cycle. These are:

　A. Compression, expansion, exhaust, and intake.
　B. Intake, compression, power, and exhaust.
　C. Induction, compression, power, and exhaust.

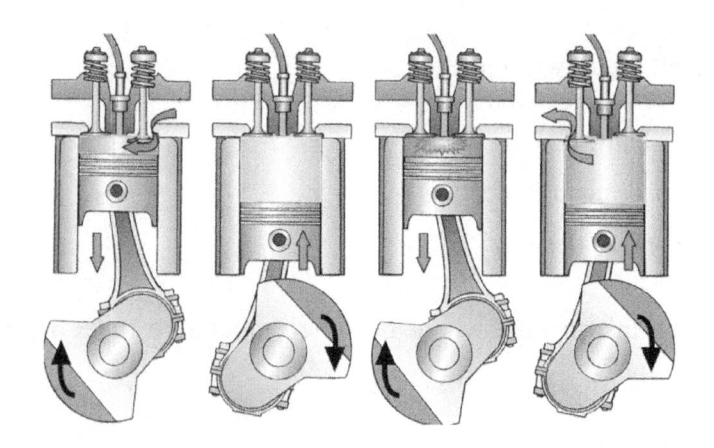

53.- The second stage of the Otto Cycle begins with the piston:

    A. Positioned at Top Dead Center (TDC).

    B. Positioned at Bottom Dead Center (BDC).

    C. The position is indifferent for this stage.

54.- In which of the four strokes of the Otto Cycle are the intake and exhaust valves completely closed??

    A. Intake and exhaust.

    B. Compression.

    C. Only intake.

55.- Regarding the propulsion system, the pilot can control it by operating the:

    A. Aceleración. Hélice. Mezcla.

    B. Hélice. Mezcla.

    C. Aceleración. Hélice.

56.- The meteorological conditions that can produce carburetor icing are:
- A. Relative humidity and temperature below 30°C.
- B. Water vapor in the air and temperature below 0°C.
- C. Relative humidity, variable cloudiness, and precipitation.

Remember! The two factors that favor the formation of carburetor ice are relative humidity and low temperatures.

57.- Extremely high engine temperature would cause:
- A. No effect on aircraft engines.
- B. Loss of power, excessive oil consumption, and possible internal damage.
- C. Damage to cylinder ducts.

58.- In an aircraft equipped with fuel pumps, when would the electric auxiliary fuel pump be used?
- A. Only when operating the aircraft in aerobatic flight conditions.
- B. Parallel to the traditional pump to anticipate possible failure.
- C. In case the mechanical fuel pump fails.

Remember! The operation of the electric fuel pump depends on the procedure for each aircraft, but it is good practice to use it during critical phases of flight as a precaution.

59.- The purpose of "leaning" the fuel mixture with altitude is:

A. To increase the amount of fuel in the mixture to compensate for the reduced amount of air.
B. To reduce the amount of fuel in the mixture to compensate for the increased air density.
C. To reduce the fuel flow to compensate for the reduction in air density.

Remember! Air density decreases with altitude, and it is necessary to adjust the fuel mixture to compensate for this difference.

60.- The use of carburetor heat tends to:

A. Increase engine performance.
B. Reduce engine performance.
C. Have no effect on engine performance.

Remember! Carburetor heat is estimated to reduce engine performance by approximately 9%.

61.- High engine oil temperature would be caused by:

A. Aircraft operated with an excessively rich mixture.
B. Extremely low oil level.
C. Extremely high outside temperature.

62.- It is recommended to fill the fuel tanks after the last flight of the day in order to:
   A. Eliminate air from the tanks and prevent vapor condensation.
   B. Prevent fuel contamination in case of rain.
   C. Eliminate remaining space in the tank to avoid debris.

Communications and flight information

When operating an aircraft within controlled airspace, there are certain communication protocols, as well as flight assistance services.

63.- Upon receiving ATIS information, you obtain:
   A. Daily meteorological and airport information.
   B. Hourly meteorological and airport information.
   C. Air traffic service and radar information.

Remember! The format of an ATIS is similar to a METAR and is issued every 60 minutes.

64.- While flying in controlled airspace on a heading of 090°, the controller reports traffic at your three o'clock position, 2 NM, heading north. The pilot should look towards:

   A. North.
   B. South.
   C. West.

65.- When using the transponder and changing its codes as required by air traffic control, the pilot should avoid selecting the following codes by mistake:

    A. 2200, 7500, 9900.
    B. 7600, 7700, 2000.
    C. 7600, 7500, 7700.

66.- In the event of a communication failure, seeing a green light from the control tower means:

    A. The aircraft should return to the airport and land.
    B. The aircraft is cleared to land.
    C. The aircraft must give way to another aircraft taking off.

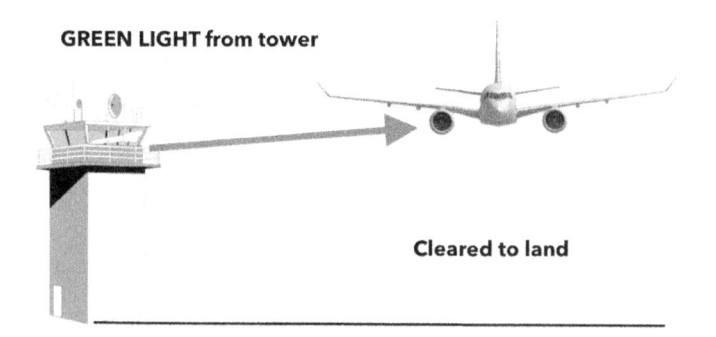

GREEN LIGHT from tower

Cleared to land

67.- In the event of a communication failure while taxiing, which of the following signals from the tower authorizes the pilot to taxi?

    A. Flashing white light.
    B. Flashing green light.
    C. Flashing blue light.

68.- When the ELT (Emergency Locator Transmitter) is activated, it transmits on the frequency:
- A. 118.0 and 118.8 MHz.
- B. 123.0 and 119.0 MHz.
- C. 121.5 and 243.0 MHz.

69.- After landing at an international airport, when should the pilot change frequency and contact ground control?En el momento en el que se le indico previamente.
- A. At the moment previously instructed.
- B. After stopping the aircraft on the landing roll.
- C. Once arrived at the parking position to complete the ground maneuver.

Remember! The frequency change from tower to ground control should occur after vacating the runway post-landing, but in all cases, the pilot must change when instructed by ATC.

Airports and Airspaces

When operating an aircraft within a specific controlled sector such as an airport or within controlled airspace, there are certain issues every pilot must know.

70.- The aircraft that will have the "right of way" over all other aircraft in airspace will be:

A. IThe hot air balloon.
B. The glider.
C. Any aircraft in an emergency.

> **Remember!** Emergency situations take precedence over any other flight condition. This grants priority to the aircraft in distress over any other aircraft.

71.- What action should pilots of two aircraft take when on a head-on collision course?La aeronave de pequeño porte sede el paso a la de gran porte.

A. The smaller aircraft gives way to the larger aircraft.
B. The larger aircraft gives way to the smaller aircraft.
C. Both aircraft make a right turn.

When two aircraft are approaching each other head-on and there is a risk of collision, each must change course to the right. Neither has the right of way.

72.- Operating in a highly congested air traffic area, before initiating any maneuver, the pilot should:

A. Check speed, altitude, and heading.
B. Visually check the entire operation area to avoid a collision.
C. Announce the next maneuver or intention to air traffic control, with which they have radio contact.

Remember! While it is very important to visually verify the flight area, any change in flight path must be reported to air traffic control whenever the aircraft is flying in a radar control zone.

73.- At an international airport, the edge lights of the taxiways are illuminated in the color:

    A. White edges and blue center.
    B. Blue edges.
    C. Green edges and white center.

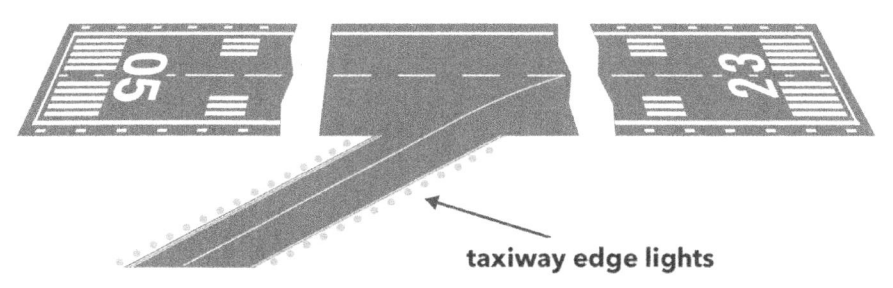

taxiway edge lights

    Remember! The edges of the taxiways are delineated with a line of blue lights.

74.- The numbers 09 and 27 on a runway indicate:
    A. That the runway orientation is 009-027.
    B. That the runway orientation is 090-027.
    C. That the runway orientation is 090-270.

    Remember! Runway numbers indicate their orientation relative to magnetic north. For example, Runway 23 is oriented at 230º.

75.- Regarding runway lighting systems, the last section of a runway will be amber and indicate:

    A. The last 900 meters remaining of the runway.

    B. The last 600 meters remaining of the runway.

    C. The last 300 meters remaining of the runway.

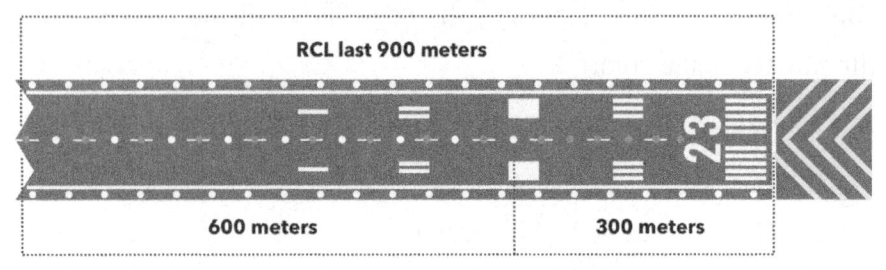

76.- On a paved runway, the touchdown zone will be identified with:

    A. An X on each side of the runway centerline if it is an uncontrolled airport.

    B. A white rectangle on each side of the runway centerline at any airport, whether domestic or international.

    C. Two parallel rectangles on each side of the runway centerline at any airport.

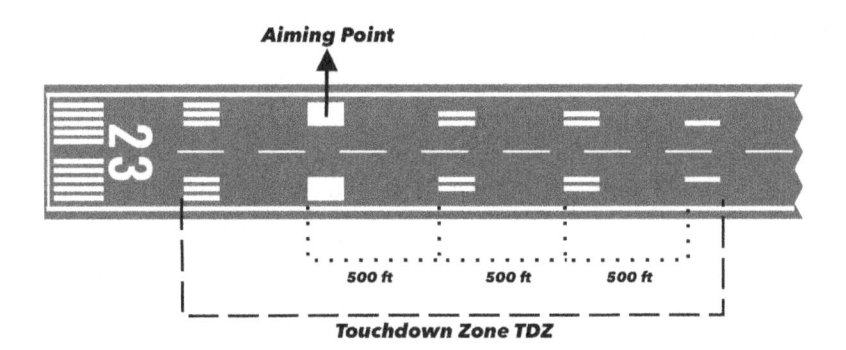

VOR Procedures

At this point, we will take additional time since developing correct VOR procedures is considered a cornerstone in every pilot's career.

77.- Understanding a VOR station as a reference point, a position is:

A. A precise point in space with a direction relative to the VOR station.
B. A precise point in space given by a radial.
C. A precise point in space given by a radial + a distance from the VOR station.

Remember! To determine an exact position relative to a VOR station, you need two values: the radial the aircraft is flying over and the distance to the VOR station.

78.- Relative to a VOR station or an NDB antenna, the following procedures are possible:

A. Approaches, Departures, Holding.
B. Approaches, Departures, DME Arc, Holding.
C. Approaches, Departures, DME Arc, Holding, Direct Flights.

79.- In an entry procedure by a given radial, the heading will be:

   A. lThe same as the desired radial.
   B. Opposite to the desired radial.
   C. The radial does not interfere with the entry heading.

Remember! In all radio procedures for entry towards a station, the heading is always opposite to the value of the entry radial, whereas in departures, these two values could be the same.

80.- Holding procedures consist of:

   A. Four stages, two of 60 seconds and two turns.
   B. Four stages, all of 60 seconds each.
   C. Two stages, one controlled and one uncontrolled.

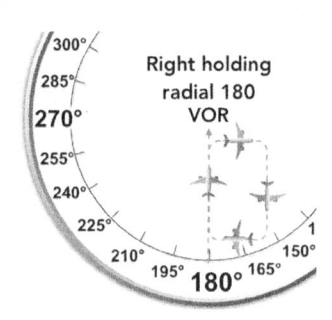

81.- The DME ARC procedure is used to:

   A. Enter or leave a position relative to an NDB.
   B. Enter or leave a position relative to a VOR.
   C. Enter or leave a position regardless of the type of antenna.

Remember! The DME Arc is used for entries and departures from any radio station, but a distance measuring equipment (DME) system is required.

82.- According to the following figure at the end of the book, in which position will the aircraft be?

A. QDM 360.
B. QDR 180.
C. Radial 340.

83.- Observing the following figure, in what position is the aircraft relative to the VOR station?

A. Radial 150.
B. Radial 170.
C. None of the above.

Using an RMI, the position should be read from the "tail" of the needle, while the direction to the station will be indicated by the tip of the needle.

84.- Observing the following figure, the VOR has 5 points on each side of the CDI, which indicate:

A. Each point, 2° deviation from the desired radial.
B. Each point, 1° deviation from the desired radial.
C. Each point, has no relation to radial deviation.

Depending on the equipment installed in the aircraft, the radial deviation indicator points represent each 2° of deviation, based on the selected radial.

85.- Observing the following figure, in what position is the aircraft relative to the VOR station?

    A. Radial 007.
    B. Radial 353.
    C. Radial 187.

86.- Observing the following figure, in what position is the aircraft relative to the NDB antenna?

    A. QDM 135.
    B. QDM 360.
    C. QDR 315.

87.- Observing the following figure, in what position is the aircraft relative to the VOR station?

    A. Radial 186.
    B. Radial 176.
    C. Radial 354.

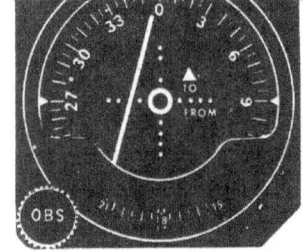

88.- Observing the following figure, to perform a hold on the 060 radial to the right, the entry will be:

A. Direct.
B. Teardrop.
C. Parallel.

Holds depicted in the upper part of the instrument use the "Teardrop" and "Parallel" entries.

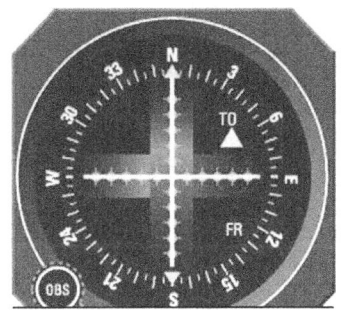

89.- Observing the previous figure, to perform a hold on the 360 radial to the left, the entry will be:

A. Direct.
B. Teardrop.
C. Parallel.

90.- The aircraft is departing on the 135 radial, and ATC instructs to enter on the 135 radial "immediately". What is the most appropriate procedure:

A. ICAO Procedure.
B. Teardrop Procedure.
C. 90-270 Procedure.

Remember! There are three course reversal procedures: Teardrop, ICAO, and 90-270. The first two require flying a specified heading for 60 seconds, while the third does not require this time of flight.

91.- The frequency band of a VOR station, in all its different classes (VOR-N and VOR-T), operates in:

A. VHF 108.0 to 118.0 MHz.
B. VHF 112.0 to 118.0 MHz.
C. VHF 118.0 to 134.40 MHz.

Remember! These radio waves are emitted at a specific frequency (from 108.00 MHz to 118.00 MHz). From the onboard equipment, the pilot can select the frequency of a specific VOR station and receive the radio signals decoded as positions.

92.- The frequency band of an NDB antenna operates in a frequency range of:a banda de frecuencia de una antena NDB opera en una banda de frecuencia de:

A. From 190 MHz to 1500 MHz in amplitude modulation (AM).
B. From 190 KHz to 1750 KHz in amplitude modulation (AM).
C. From 200 KHz to 900 KHz in amplitude modulation (AM).

Aviation Regulations

There are certain regulatory aspects that every pilot starting their first steps in aviation should know. For this purpose, there is the well-known flight regulations manual, where the most important rules to follow are detailed.

93.- Under optimal weather conditions, a VFR flight can be carried out within the following altitude or level ranges:

    A. From the surface to the transition layer.

    B. From the surface to flight level 195.

    C. At any flight level as long as 500 FT is added.

Remember! All VFR flights, whether controlled or uncontrolled, can be conducted up to 19,500 feet. For higher altitudes, flights must be conducted under IFR rules.

94.- A VFR flight must always maintain visual contact with the ground and may fly over clouds as long as they do not cover:

    A. The entire surface, preventing visual contact.

    B. More than four-eighths of the surface.

    C. More than one-third of the surface.

Remember! In all cases, VFR flights must maintain visual contact with the ground throughout the journey. If permanent visual contact with the surface is lost, the flight must be conducted under IFR.

95.- The minimum height for a VFR flight over the sea, terrain, and city will be:

    A. Sea 500FT. Terrain 500FT. City 1000FT.

    B. Sea 1000FT. Terrain 500FT. City 1000FT.

    C. Sea 500FT. Terrain 500FT. City 500FT.

96.- The weather minimums for an uncontrolled aerodrome during a VFR flight will be:

    A. Visibility 5KM and cloud ceiling at 1000FT.
    B. Visibility 2.5KM and cloud ceiling at 500FT.
    C. Visibility 5000FT and cloud ceiling at 1000MTS.

Remember! International regulations establish a minimum of 5,000 meters of horizontal visibility as the limit for VFR flights.

97.- Regarding the transition layer, it has a transition level at which the pilot must change the altimeter setting:

    A. To flight level if ascending.
    B. To flight level regardless of whether ascending or descending.
    C. To the local QNH.

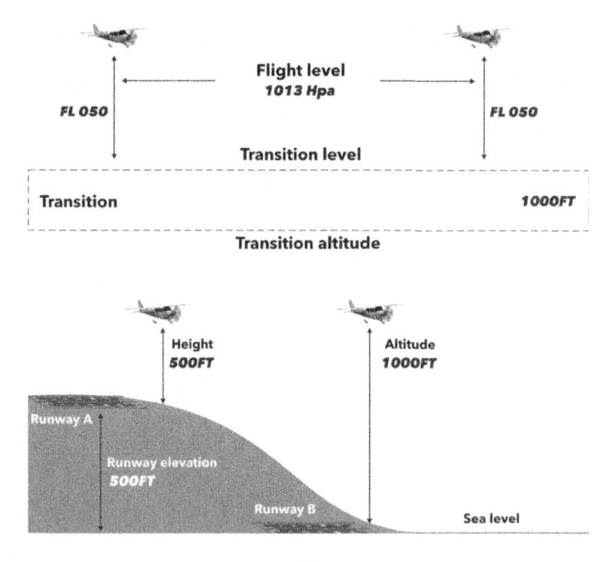

98.- The regulatory navigation lights on an aircraft are:

    A. Left light red, right light green, tail light white.

    B. Left light green, right light red, no tail light.

    C. The position of the lights is irrelevant and may vary.

99.- Regarding a flight plan, it should be filed:

    A. Before every flight, whether controlled or uncontrolled.

    B. Only for instrumental flights.

    C. For any controlled flight, whether VFR or IFR.

Remember! The flight plan form is a sworn statement of the intended flight activity. Its purpose is to regulate all controlled flights, regardless of the flight rules anticipated, and to provide flight assistance services.

100.- A flight plan form, once submitted, will be valid for:

    A. 30 minutes and then must be amended.

    B. 45 minutes and then must be amended.

    C. The flight plan never expires once submitted.

Remember! The flight plan is valid for 30 minutes from its submission. If this period expires, the pilot can submit an amendment for the delay, and it will have the same validity period again.

Flight Plan

It's important to practice filling out this form, as it involves a variety of codes and formats to remember.

101.- Complete the flight plan form using the following operational data:

For this initial flight, we will plan a short and very simple route. The flight takes place within the same TMA, from a small airport to the runway of a neighboring aeroclub just a few miles away. The aircraft is a small training aircraft, a Cessna 152, with the instructor and their student pilot onboard.

They have filled the fuel tanks to half of their maximum capacity, which provides them with approximately 2 hours and 45 minutes of autonomy. The flight will be conducted at 13:30 local time under visual flight rules, at 1500 feet.

Departure: SAZM Airport.
Destination: Private aerodrome TEC.

Below is a blank flight plan. Try to complete it, and then we will finalize it together on the next page.

U.S Department of Transportation
Federal Aviation Administration

International Flight Plan

PRIORITY   ADDRESSEE(S)
<=FF
                                                                <=

FILING TIME   ORIGINATOR
                                        <=

SPECIFIC IDENTIFICATION OF ADDRESSEE(S) AND / OR ORIGINATOR

3 MESSAGE TYPE   7 AIRCRAFT IDENTIFICATION   8 FLIGHT RULES   TYPE OF FLIGHT
<=(FPL   —   —   —   <=
9 NUMBER   TYPE OF AIRCRAFT   WAKE TURBULENCE CAT.   10 EQUIPMENT
—   /   —   /   <=
13 DEPARTURE AERODROME   TIME
—   <=
15 CRUISING SPEED   LEVEL   ROUTE
—

                                                                <=

                    TOTAL EET
16 DESTINATION AERODROME   HR   MIN   ALTN AERODROME   2ND ALTN AERODROME
                                                                            <=
18 OTHER INFORMATION
—

                                                                <=

SUPPLEMENTARY INFORMATION (NOT TO BE TRANSMITTED IN FPL MESSAGES)
19 ENDURANCE   EMERGENCY RADIO
    HR   MIN   PERSONS ON BOARD   UHF   VHF   ELBA
—E/   P/   R/ ☐   ☐   ☐

    SURVIVAL EQUIPMENT   JACKETS
        POLAR DESERT MARITIME JUNGLE   LIGHT FLUORES UHF VHF
    ☐ / ☐   ☐   ☐   ☐   ☐ / ☐   ☐   ☐   ☐
    DINGHIES
    NUMBER CAPACITY COVER   COLOR
D/   <=
    AIRCRAFT COLOR AND MARKINGS
A/
    REMARKS
N/   <=
    PILOT-IN-COMMAND
C/   )<=
        FILED BY   ACCEPTED BY   ADDITIONAL INFORMATION

International Flight Plan

U S Department of Transportation / Federal Aviation Administration

PRIORITY ADDRESSEE(S)
<=FF

FILING TIME ORIGINATOR <=

SPECIFIC IDENTIFICATION OF ADDRESSEE(S) AND / OR ORIGINATOR

3 MESSAGE TYPE: <=(FPL
7 AIRCRAFT IDENTIFICATION: — LV-TEC
8 FLIGHT RULES: — V
TYPE OF FLIGHT: — G <=

9 NUMBER: —
TYPE OF AIRCRAFT: C152
WAKE TURBULENCE CAT.: / L
10 EQUIPMENT: — S / C <=

13 DEPARTURE AERODROME: — SAZM
TIME: 1630 <=

15 CRUISING SPEED: — N0100
LEVEL: A015
ROUTE: DCT ZZZZ

16 DESTINATION AERODROME: ZZZZ
TOTAL EET HR MIN: 0020
ALTN AERODROME: SAZM
2ND ALTN AERODROME: <=

18 OTHER INFORMATION:
— DEST/ TEC AERODROME
OPR/ FC FLIGHT SCHOOL
<=

SUPPLEMENTARY INFORMATION (NOT TO BE TRANSMITTED IN FPL MESSAGES)

19 ENDURANCE HR MIN: —E/ 0245
PERSONS ON BOARD: P/ 002
EMERGENCY RADIO: R/ [X] UHF [ ] VHF [X] ELBA

SURVIVAL EQUIPMENT: [X] / POLAR [X] DESERT [X] MARITIME [X] JUNGLE [X]
JACKETS: [X] / LIGHT [X] FLUORES [X] UH [X] VHF [X]

DINGHIES NUMBER CAPACITY COVER: [X]
COLOR: <=

AIRCRAFT COLOR AND MARKINGS
A/ WHITE AIRCRAFT WITH BLUE MARKS

REMARKS
[X] / <=

PILOT-IN-COMMAND
C/ JUAN PEREZ)<=

FILED BY   ACCEPTED BY   ADDITIONAL INFORMATION

Let's analyze this example together. In the first section, the initial field to be completed is the aircraft registration (using a fictional example for illustration). Remember, it's not necessary to fill in all spaces within a cell. If there are excess spaces, they can remain blank but should never be crossed out or marked.

```
3 MESSAGE TYPE     7 AIRCRAFT IDENTIFICATION        8 FLIGHT RULES    TYPE OF FLIGHT
<=(FPL              — |L,V,-,T,E,C|                  — |V|             — |G|          <=
9 NUMBER           TYPE OF AIRCRAFT    WAKE TURBULENCE CAT.           10 EQUIPMENT
—|  |               |C,1,5,2|              / |L|                  — |      S / C |    <=
       13 DEPARTURE AERODROME           TIME
          — |S,A,Z,M|             |  ,  1,6,3,0|  <=
       15 CRUISING SPEED    LEVEL        ROUTE
       —|N,0,1,0,0|      |A,0,1,5|   |DCT ZZZZ|
```

The next field requires filling in the flight rules with the first code to remember. According to the flight information, this will be conducted under Visual Flight Rules (VFR). In this case, the code for a VFR flight is the letter V.

The subsequent field requires declaring the type of flight, which, for this case, will be general aviation, with the code G.

Below and to the left, following the filling order from right to left and from top to bottom, is the aircraft type field. The flight information indicates that the aircraft is a Cessna model 152. The code for this type of aircraft is C152. Considering its maximum takeoff weight is below 7,000 kilograms, it will be categorized as light turbulence wake, to be completed with the letter L in the next field to the right.

In the field on the right, a standard equipment declaration is made with the letter S, and a Mode C transponder is indicated with the letter C following the separator bar.

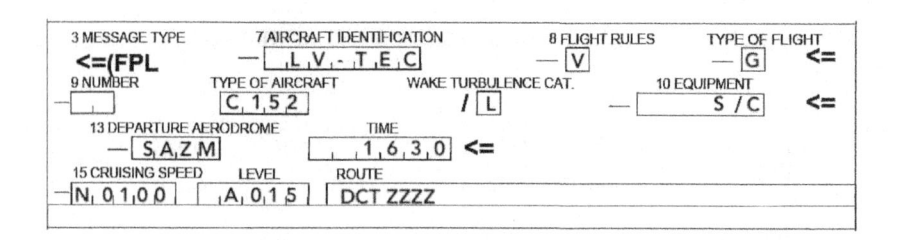

```
3 MESSAGE TYPE      7 AIRCRAFT IDENTIFICATION           8 FLIGHT RULES      TYPE OF FLIGHT
  <=(FPL              — L,V,-,T,E,C                      — V                 — G         <=
9 NUMBER            TYPE OF AIRCRAFT    WAKE TURBULENCE CAT.              10 EQUIPMENT
  —                   C,1,5,2                  / L                         —   S / C     <=
13 DEPARTURE AERODROME            TIME
  — S,A,Z M                       1,6,3,0  <=
15 CRUISING SPEED   LEVEL       ROUTE
— N, 0,1,0,0        A, 0,1,5    DCT ZZZZ
```

Below and to the left, the field for the departure aerodrome is located. In this case, it is the Mar del Plata Airport in Argentina. Its international ICAO designator is SAZM, which should be entered in the blank spaces of this field.

To the right, the field for the estimated start-up time is located. According to the flight information, the estimated start-up is at 13:30 local time. However, in this field, the time must be entered in UTC. For Argentina, the UTC time is local time + 3 hours. Therefore, you should enter 16:30.

In the lower section of this first part, there are the last three fields. On the left side, the field for the speed. In this case, the pilot has declared a cruising speed of 100 knots. To the right, the field for altitude or flight level. Here, the flight is planned to cruise at an altitude of 1,500 feet. For this information, the code is the letter A (for altitude) and the value expressed in hundreds of feet; 15 hundreds equate to 1,500 feet.

Finally, in the last field on the right, the route section. For this flight, the segment is quite short and without intermediate points. The flight will proceed directly from the origin to the destination aerodrome, which, not having an ICAO designator, is established as ZZZZ. Subsequently, its name should be clarified.

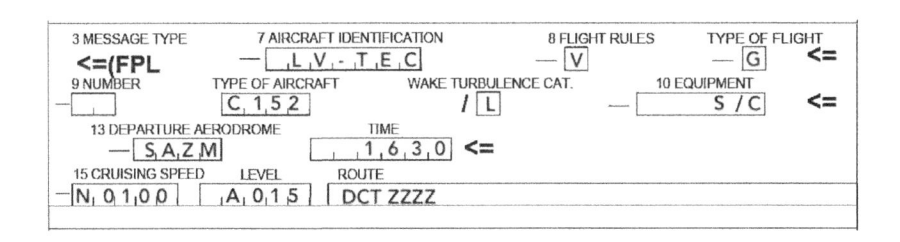

3 MESSAGE TYPE	7 AIRCRAFT IDENTIFICATION	8 FLIGHT RULES	TYPE OF FLIGHT
<=(FPL	— L V - T E C	— V	— G <=
9 NUMBER	TYPE OF AIRCRAFT WAKE TURBULENCE CAT.		10 EQUIPMENT
—	C 1 5 2 / L	—	S / C <=
13 DEPARTURE AERODROME	TIME		
— S A Z M	1 6 3 0 <=		
15 CRUISING SPEED LEVEL	ROUTE		
— N 0 1 0 0 A 0 1 5	DCT ZZZZ		

In the second section of the flight plan, the first field requires declaring the ICAO designator of the destination. In this case, as the aerodrome does not have one, the code ZZZZ is used, and the location will be specified later.

16 DESTINATION AERODROME	TOTAL EET HR MIN	ALTN AERODROME	2ND ALTN AERODROME
Z Z Z Z	0 0 2 0	S A Z M	<=
18 OTHER INFORMATION			
— DEST/ TEC AERODROME			
OPR/ FC FLIGHT SCHOOL			

The total flight time, from takeoff to landing, is 20 minutes. In the EET (Estimated Elapsed Time) section, this is declared as 00 hours and 20 minutes. To the right, the fields for alternate aerodromes are provided. As this is a VFR flight, only one alternate aerodrome is required. In this case, the same aerodrome as the origin is declared as the alternate.

Finally, in the "Other Information" field, it is necessary to specify the destination aerodrome and the operator of the aircraft. Remember that the operator of the aircraft is not the pilot but the legal responsible entity and owner of the aircraft.

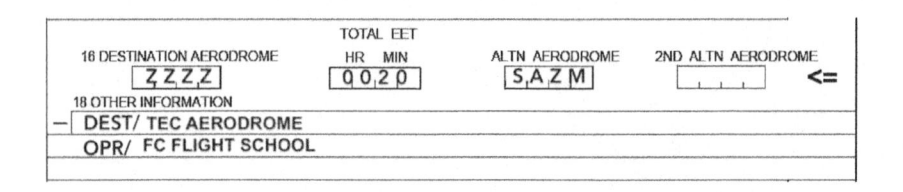

```
16 DESTINATION AERODROME   TOTAL EET        ALTN AERODROME   2ND ALTN AERODROME
       ZZZZ                HR  MIN              SAZM
                           0020                                                    <=
18 OTHER INFORMATION
—  DEST/ TEC AERODROME
   OPR/ FC FLIGHT SCHOOL
```

In this case, the destination aerodrome is declared with the code DEST/ and the name of the location (aerodrome TEC); and the operator is declared with the code OPR/ and the name of the flight academy.

Moving on to the supplementary information section, it starts with the field for the aircraft's endurance. This value is expressed in hours and minutes of flight. According to the flight information, the aircraft has an endurance of 2 hours and 45 minutes, and there will be 2 occupants on board, which is entered in the passengers field.

```
      SUPPLEMENTARY INFORMATION (NOT TO BE TRANSMITTED IN FPL MESSAGES)
19    ENDURANCE                                          EMERGENCY RADIO
      HR  MIN        PERSONS ON BOARD              UHF    VHF    ELBA
—E/  0 2 4 5    P/  0 0 2                      R/ X      □      X
    SURVIVAL EQUIPMENT              JACKETS
         POLAR DESERT MARITIME JUNGLE      LIGHT FLUORES  UH    VHF
     X/  X    X       X       X         X/  X       X    X     X
    DINGHIES
    NUMBER CAPACITY COVER      COLOR
 D/  □   □       □          □                     <=
    AIRCRAFT COLOR AND MARKINGS
 A/  WHITE AIRCRAFT WITH BLUE MARKS
    REMARKS
 X/                                                              <=
    PILOT-IN-COMMAND
 C/  JUAN PEREZ                            )<=
        FILED BY           ACCEPTED BY           ADDITIONAL INFORMATION
```

54

Regarding emergency, radio, and survival equipment, the aircraft is equipped with only a VHF radio. Considering this, the pilot should cross out all other fields related to emergency and survival equipment, except for the VHF radio field.

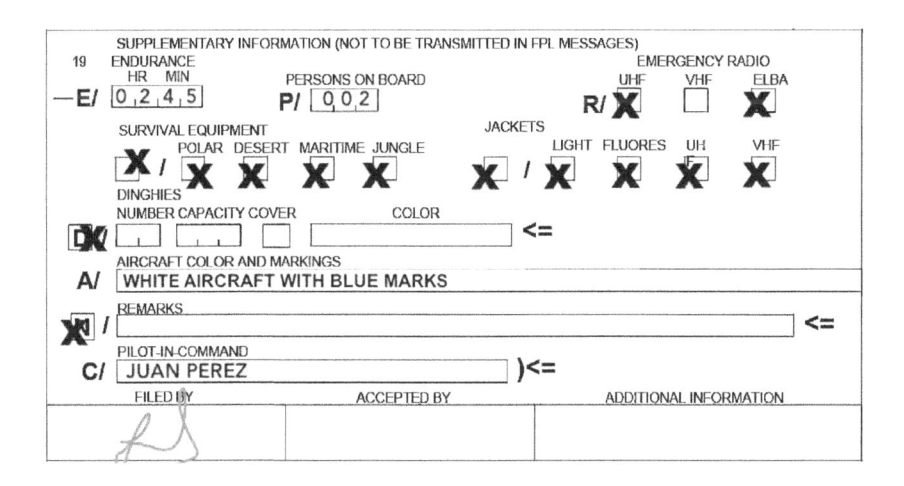

```
19  SUPPLEMENTARY INFORMATION (NOT TO BE TRANSMITTED IN FPL MESSAGES)
    ENDURANCE                                        EMERGENCY RADIO
    HR  MIN       PERSONS ON BOARD              UHF    VHF    ELBA
-E/ 0,2,4,5    P/ 0,0,2                      R/ X      ☐      X
    SURVIVAL EQUIPMENT              JACKETS
    POLAR DESERT MARITIME JUNGLE           LIGHT FLUORES UH   VHF
    X / X   X   X   X              X / X    X    X    X
    DINGHIES
    NUMBER CAPACITY COVER    COLOR
 X  ☐    ☐      ☐     ☐                  <=
    AIRCRAFT COLOR AND MARKINGS
 A/ WHITE AIRCRAFT WITH BLUE MARKS
    REMARKS
 X/                                                         <=
    PILOT-IN-COMMAND
 C/ JUAN PEREZ                        )<=
    FILED BY          ACCEPTED BY        ADDITIONAL INFORMATION
```

The last three rows declare the aircraft's color and markings. In this case, the aircraft is white with black and blue stripes. No observations related to emergency and survival equipment information are declared.

The name of the pilot in command is declared, along with their signature, as they presented the flight plan to the airport's aviation authority.

Have you managed to complete it correctly? If there were any errors, don't worry—it's just a matter of practice.

102.- Complete the flight plan form using the following operational data:

This example plans a local coastal flight for sightseeing purposes. The aircraft is a Piper Seneca V (PA-34), registration N00123, operated by two pilots and carrying 4 tourists. The cruising speed will be 100 knots, and the planned altitude is 2000 feet, with a low-altitude overflight along the coast. The flight will depart from an uncontrolled aerodrome (a local aeroclub) at 12:00 local time and will last about 20 minutes. One of the pilots has personally delivered the flight plan to the operations office at the airport earlier that morning.

The destination aerodrome will be the same aeroclub from which the flight departed. Alternatively, the closest airport is the international airport of Montevideo, Uruguay, ICAO designator SUMU. The aircraft has fuel tanks filled to half capacity, providing a range of 3.5 hours of flight.

The aircraft is light blue with blue stripes, equipped with VHF radio and an ELT (Emergency Locator Transmitter) as its emergency beacon. Since this is a sea overflight, the aircraft is equipped with 6 life vests. These vests have reflective lights and colors but do not include any beacon systems.

Note: In Uruguay, as in Argentina, the local time is UTC + 3 hours.

International Flight Plan

U S Department of Transportation
Federal Aviation Administration

PRIORITY   ADDRESSEE(S)
<=FF

⇐

FILING TIME   ORIGINATOR
⇐

SPECIFIC IDENTIFICATION OF ADDRESSEE(S) AND / OR ORIGINATOR

3 MESSAGE TYPE   7 AIRCRAFT IDENTIFICATION   8 FLIGHT RULES   TYPE OF FLIGHT
<=(FPL   —   —   —   ⇐
9 NUMBER   TYPE OF AIRCRAFT   WAKE TURBULENCE CAT.   10 EQUIPMENT
—   /   —   /   ⇐
13 DEPARTURE AERODROME   TIME
—   ⇐
15 CRUISING SPEED   LEVEL   ROUTE
—

⇐

TOTAL EET
16 DESTINATION AERODROME   HR MIN   ALTN AERODROME   2ND ALTN AERODROME
⇐
18 OTHER INFORMATION
—

⇐

SUPPLEMENTARY INFORMATION (NOT TO BE TRANSMITTED IN FPL MESSAGES)
19   ENDURANCE   EMERGENCY RADIO
      HR MIN   UHF   VHF   ELBA
—E/   ☐   PERSONS ON BOARD   R/ ☐   ☐   ☐
              P/

       SURVIVAL EQUIPMENT   JACKETS
              POLAR DESERT MARITIME JUNGLE   LIGHT FLUORES UH   VHF
              ☐ / ☐   ☐   ☐   ☐ / ☐   ☐F   ☐

       DINGHIES
       NUMBER CAPACITY COVER   COLOR
D/   ☐   ☐   ⇐

       AIRCRAFT COLOR AND MARKINGS
A/

       REMARKS
N/   ⇐

       PILOT-IN-COMMAND
C/   )⇐

       FILED BY   ACCEPTED BY   ADDITIONAL INFORMATION

57

U.S Department of Transportation
Federal Aviation Administration

International Flight Plan

PRIORITY   ADDRESSEE(S)
<=FF

FILING TIME   ORIGINATOR   <=

SPECIFIC IDENTIFICATION OF ADDRESSEE(S) AND / OR ORIGINATOR

3 MESSAGE TYPE   7 AIRCRAFT IDENTIFICATION   8 FLIGHT RULES   TYPE OF FLIGHT
<=(FPL   — N 0 0 1 2 3   — V   — G   <=

9 NUMBER   TYPE OF AIRCRAFT   WAKE TURBULENCE CAT.   10 EQUIPMENT
—   P A 3 4   / L   — S / C   <=

13 DEPARTURE AERODROME   TIME
— Z Z Z Z   1 5 0 0   <=

15 CRUISING SPEED   LEVEL   ROUTE
— N 1 0 0   A 0 2 0   LOCAL FLIGHT RADIALS 360 AND 180
20NM FROM SUMU VOR

<=

                      TOTAL EET
16 DESTINATION AERODROME   HR MIN   ALTN AERODROME   2ND ALTN AERODROME
Z Z Z Z   0 0 2 0   S U M U   <=

18 OTHER INFORMATION
— DEP/ MPP AERODROME
DEST/ MPP AERODROME

<=

SUPPLEMENTARY INFORMATION (NOT TO BE TRANSMITTED IN FPL MESSAGES)
19   ENDURANCE   EMERGENCY RADIO
     HR MIN   PERSONS ON BOARD   UHF   VHF   ELBA
— E/ 0 3 3 0   P/ 0 0 6   R/ X   □   □

SURVIVAL EQUIPMENT   JACKETS
   POLAR DESERT MARITIME JUNGLE   LIGHT FLUORES UH VHF
X / X   X   X   X   □ / □   □   X   X

DINGHIES
NUMBER CAPACITY COVER   COLOR
X̷D /   □   □   □   □   <=

AIRCRAFT COLOR AND MARKINGS
A/ SKY BLUE AIRCRAFT WITH BLUE MARKINGS

REMARKS
X̷N /   <=

PILOT-IN-COMMAND
C/ Juan Marquez   )<=

FILED BY   ACCEPTED BY   ADDITIONAL INFORMATION

Julian Ramón

The flight plan begins with the registration number clarification, followed by the letter V for Visual Flight Rules, and the letter G for General Flight. The aircraft type is PA34, with a light turbulence category and standard equipment with transponder mode C.

The origin aerodrome is a local aeroclub without an OACI designator; therefore, ZZZZ is declared, and the location name will be provided later. The flight departs at 15:00 UTC (12:00 local time). A speed of 100 knots and an altitude of 2000 feet are declared, using the letter A and the value in hundreds of feet (20 hundreds is equal to 2000 feet). The route is described as a local flight, covering the eastern sector of the aerodrome within a 20-mile radius.

3 MESSAGE TYPE	7 AIRCRAFT IDENTIFICATION	8 FLIGHT RULES	TYPE OF FLIGHT
<=(FPL	— N 0 0 1 2 3	— V	— G <=
9 NUMBER	TYPE OF AIRCRAFT WAKE TURBULENCE CAT.	10 EQUIPMENT	
—	P A 3 4 / L	— S/C <=	
13 DEPARTURE AERODROME	TIME		
— Z Z Z Z	1 5 0 0 <=		
15 CRUISING SPEED LEVEL	ROUTE		
N 1 0 0 A 0 2 0	LOCAL FLIGHT RADIALS 360 AND 180		
	20NM FROM SUMU VOR		

The second section declares the destination aerodrome as ZZZZ, as it is the same origin aerodrome, which does not have an OACI designator.

	TOTAL EET		
16 DESTINATION AERODROME	HR MIN	ALTN AERODROME	2ND ALTN AERODROME
Z Z Z Z	0 0 2 0	S U M U	<=
18 OTHER INFORMATION			
— DEP/ MPP AERODROME			
DEST/ MPP AERODROME			

The flight time is 20 minutes, and the alternate destination is Montevideo International Airport, Uruguay. In item 18, the names of the origin and destination are clarified in plain text with the codes DEP and DEST, respectively.

The final section declares a flight endurance of 3 hours and 30 minutes, with 6 persons onboard the aircraft, equipped with a VHF radio, an ELT, and life vests with a light system but no radio equipment. Finally, the aircraft description and signatures of both pilots are included as appropriate.

SUPPLEMENTARY INFORMATION (NOT TO BE TRANSMITTED IN FPL MESSAGES)	
19 ENDURANCE HR MIN — E/ 0 3 3 0	PERSONS ON BOARD P/ 0 0 6	EMERGENCY RADIO UHF VHF ELBA R/ X ☐ ☐
SURVIVAL EQUIPMENT POLAR DESERT MARITIME JUNGLE X / X X X X	JACKETS LIGHT FLUORES UH VHF ☐ / ☐ ☐ X X
DINGHIES NUMBER CAPACITY COVER COLOR D/ ☐☐ ☐☐☐ ☐ <=	
AIRCRAFT COLOR AND MARKINGS A/ SKY BLUE AIRCRAFT WITH BLUE MARKINGS	
REMARKS N/	<=
PILOT-IN-COMMAND C/ Juan Marquez)<=	
FILED BY Julian Ramón	ACCEPTED BY	ADDITIONAL INFORMATION

103.- Complete the flight plan form using the following operational data:

In this example, we will plan a flight from an airport on September 14th to a ranch in the countryside and return the next day, departing from the same ranch to the original airport. The pilot will file two flight plans, one for the flight on the same day and another for the return flight the following day.

The aircraft is a Cessna 182, registration MN-0510. The pilot is carrying three additional passengers who will stay at the ranch for a week. The aircraft is equipped with standard equipment and a Mode C transponder. It will depart with full fuel tanks and has a range of 5 hours.

The departure airport is Córdoba City (SACO), Argentina. The destination is a ranch located 45 NM south of the airport, on radial CBA170. The aircraft's cruising speed is 120 knots, and it is planned to fly at an altitude of 1500 meters on a direct route to the runway of the ranch named "Estancia Olga."

The flight time is 25 minutes, with a planned departure at 08:00 local time. As it is a VFR flight, the only alternate destination is the departure airport. The pilot in command is the aircraft owner, Captain Jose Enrique, who will file both flight plans. The aircraft is white with red and is equipped with only a VHF radio.

The return flight is planned directly to Córdoba Airport (SACO), under the same conditions, but considering departure at 12:00 local time.

International Flight Plan

U S Department of Transportation
Federal Aviation Administration

PRIORITY ADDRESSEE(S)
<=FF

FILING TIME   ORIGINATOR   <=

SPECIFIC IDENTIFICATION OF ADDRESSEE(S) AND / OR ORIGINATOR

3 MESSAGE TYPE   7 AIRCRAFT IDENTIFICATION   8 FLIGHT RULES   TYPE OF FLIGHT
<=(FPL   —   —   —   <=
9 NUMBER   TYPE OF AIRCRAFT   WAKE TURBULENCE CAT.   10 EQUIPMENT
—   /   —   /   <=
13 DEPARTURE AERODROME   TIME
—   <=
15 CRUISING SPEED   LEVEL   ROUTE
—

TOTAL EET
16 DESTINATION AERODROME   HR MIN   ALTN AERODROME   2ND ALTN AERODROME   <=
18 OTHER INFORMATION
—

<=

SUPPLEMENTARY INFORMATION (NOT TO BE TRANSMITTED IN FPL MESSAGES)
19 ENDURANCE   EMERGENCY RADIO
       HR   MIN   PERSONS ON BOARD   UHF   VHF   ELBA
—**E/**   **P/**   **R/**
    SURVIVAL EQUIPMENT   JACKETS
        POLAR DESERT MARITIME JUNGLE   LIGHT FLUORES UHF VHF
          /   /
    DINGHIES
    NUMBER CAPACITY COVER   COLOR
D/   <=
    AIRCRAFT COLOR AND MARKINGS
A/
    REMARKS
N/   <=
    PILOT-IN-COMMAND
C/   )<=

FILED BY   ACCEPTED BY   ADDITIONAL INFORMATION

International Flight Plan

U S Department of Transportation
Federal Aviation Administration

PRIORITY ADDRESSEE(S)
<=FF
                                                                          <=

FILING TIME   ORIGINATOR
                                          <=

SPECIFIC IDENTIFICATION OF ADDRESSEE(S) AND / OR ORIGINATOR

3 MESSAGE TYPE   7 AIRCRAFT IDENTIFICATION   8 FLIGHT RULES   TYPE OF FLIGHT
<=(FPL   —   — ☐   — ☐   <=
9 NUMBER   TYPE OF AIRCRAFT   WAKE TURBULENCE CAT.   10 EQUIPMENT
—   /☐   —   /   <=
13 DEPARTURE AERODROME   TIME
—   <=
15 CRUISING SPEED   LEVEL   ROUTE
—

                                                                          <=
                          TOTAL EET
16 DESTINATION AERODROME   HR MIN   ALTN AERODROME   2ND ALTN AERODROME
                                                                          <=
18 OTHER INFORMATION
—

                                                                          <=
        SUPPLEMENTARY INFORMATION (NOT TO BE TRANSMITTED IN FPL MESSAGES)
19   ENDURANCE   EMERGENCY RADIO
     HR MIN   UHF   VHF   ELBA
—E/   PERSONS ON BOARD   ☐   ☐   ☐
               P/   R/

     SURVIVAL EQUIPMENT   JACKETS
       POLAR DESERT MARITIME JUNGLE   LIGHT FLUORES UHF   VHF
       ☐ / ☐   ☐   ☐   ☐ / ☐   ☐   ☐
     DINGHIES
     NUMBER CAPACITY COVER   COLOR
D/   <=
     AIRCRAFT COLOR AND MARKINGS
A/
     REMARKS
N/   <=
     PILOT-IN-COMMAND
C/   )<=

        FILED BY   ACCEPTED BY   ADDITIONAL INFORMATION

U.S Department of Transportation
Federal Aviation Administration

International Flight Plan

PRIORITY   ADDRESSEE(S)
<=FF

<=

FILING TIME   ORIGINATOR   <=

SPECIFIC IDENTIFICATION OF ADDRESSEE(S) AND / OR ORIGINATOR

3 MESSAGE TYPE   7 AIRCRAFT IDENTIFICATION   8 FLIGHT RULES   TYPE OF FLIGHT
<=(FPL   — MN0510   — V   — G   <=
9 NUMBER   TYPE OF AIRCRAFT   WAKE TURBULENCE CAT.   10 EQUIPMENT
—   C182   / L   — S/C   <=
13 DEPARTURE AERODROME   TIME
— SACO   1100   <=
15 CRUISING SPEED   LEVEL   ROUTE
— N120   M0150   DCT CBA17045

<=

TOTAL EET
16 DESTINATION AERODROME   HR MIN   ALTN AERODROME   2ND ALTN AERODROME
ZZZZ   0025   SACO   <=
18 OTHER INFORMATION
— DEST/ ESTANCIA "OLGA" CBA170 45 NM

<=

SUPPLEMENTARY INFORMATION (NOT TO BE TRANSMITTED IN FPL MESSAGES)
19   ENDURANCE   EMERGENCY RADIO
HR MIN   PERSONS ON BOARD   UHF   VHF   ELBA
— E/ 0500   P/ 004   R/ X   ☐   X
SURVIVAL EQUIPMENT   JACKETS
POLAR DESERT MARITIME JUNGLE   LIGHT FLUORES UH VHF
X / X   X   X   X   X / X   X   X   X
DINGHIES
NUMBER CAPACITY COVER   COLOR
X D /   <=
AIRCRAFT COLOR AND MARKINGS
A/ WHITE AND RED AIRCRAFT
REMARKS
X N /   <=
PILOT-IN-COMMAND
C/ JOSE ENRIQUE   )<=
FILED BY   ACCEPTED BY   ADDITIONAL INFORMATION
JOSE ENRIQUE

International Flight Plan

U.S Department of Transportation
Federal Aviation Administration

PRIORITY `<=FF`
ADDRESSEE(S)

FILING TIME **ORIGINATOR** `<=`

SPECIFIC IDENTIFICATION OF ADDRESSEE(S) AND / OR ORIGINATOR

3 MESSAGE TYPE	7 AIRCRAFT IDENTIFICATION	8 FLIGHT RULES	TYPE OF FLIGHT
`<=(FPL`	MN0510	V	G `<=`

9 NUMBER	TYPE OF AIRCRAFT	WAKE TURBULENCE CAT.	10 EQUIPMENT
	C182	/L	S/C `<=`

13 DEPARTURE AERODROME	TIME
ZZZZ	1500 `<=`

15 CRUISING SPEED	LEVEL	ROUTE
N120	M0150	DCT SACO

`<=`

16 DESTINATION AERODROME	TOTAL EET HR MIN	ALTN AERODROME	2ND ALTN AERODROME
SACO	0025	ZZZZ	`<=`

18 OTHER INFORMATION
DEPT/ ESTANCIA "OLGA" CBA170 45 NM
ALTN/ ESTANCIA "OLGA" CBA170 45 NM
DOF/ FLIGHT DATE SEPTEMBER 15 `<=`

SUPPLEMENTARY INFORMATION (NOT TO BE TRANSMITTED IN FPL MESSAGES)

19 ENDURANCE HR MIN
E/ 0500 P/ 001 **PERSONS ON BOARD**

EMERGENCY RADIO UHF VHF ELBA
R/ X ☐ X

SURVIVAL EQUIPMENT POLAR DESERT MARITIME JUNGLE
X / X X X X

JACKETS LIGHT FLUORES UHF VHF
X / X X X X

DINGHIES NUMBER CAPACITY COVER COLOR
X̶D / ☐ ☐ ☐ `<=`

AIRCRAFT COLOR AND MARKINGS
A/ WHITE AND RED AIRCRAFT

REMARKS
X̶N / `<=`

PILOT-IN-COMMAND
C/ JOSE ENRIQUE)`<=`

FILED BY	ACCEPTED BY	ADDITIONAL INFORMATION
JOSE ENRIQUE		

The first section of the flight plan starts with the aircraft registration, followed by visual flight rules (VFR) and the general type of flight. Below, it specifies the aircraft type designation, light turbulence category, and standard equipment with Mode Charlie transponder.

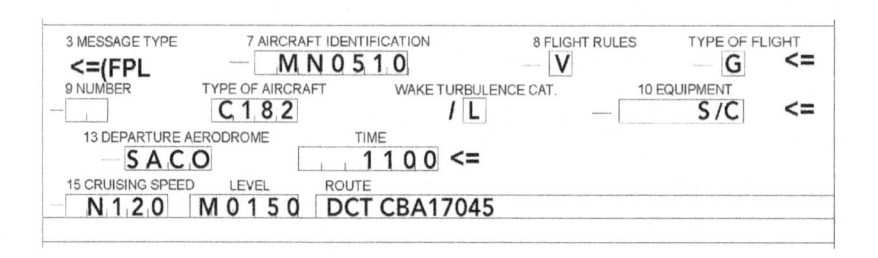

```
3 MESSAGE TYPE      7 AIRCRAFT IDENTIFICATION    8 FLIGHT RULES    TYPE OF FLIGHT
<=(FPL              — M N 0 5 1 0              —  V              — G         <=
9 NUMBER            TYPE OF AIRCRAFT    WAKE TURBULENCE CAT.     10 EQUIPMENT
—                   C 1 8 2             / L                    —  S /C       <=
13 DEPARTURE AERODROME        TIME
— S A C O                     1 1 0 0  <=
15 CRUISING SPEED   LEVEL       ROUTE
— N 1 2 0           M 0 1 5 0   DCT CBA17045
```

According to the flight information, departure from Córdoba Airport is planned for 08:00 local time. In UTC, this is declared as 11:00. Remember that in Argentina, UTC is local time + 3 hours. The pilot has declared a speed of 120 knots and an altitude of 1500 meters. Note that when declaring in this unit of measure, it must be expressed in tens of meters. In this case, "M" denotes 150 tens, which equals 1500 meters. The route description indicates a direct flight to a specified position, namely the location at 45 nautical miles on radial 170 from the Córdoba VOR. The second section continues with the declaration of the destination.

In this case, it is declared as ZZZZ, and the location will be specified in plain text later.

```
                              TOTAL EET
16 DESTINATION AERODROME      HR  MIN      ALTN AERODROME    2ND ALTN AERODROME
    Z Z Z Z                   0 0 2 5        S A C O                              <=
18 OTHER INFORMATION
    DEST/ ESTANCIA "OLGA" CBA170 45 NM
```

The provided flight time is 25 minutes, from takeoff to landing, and the airport of origin has been declared as the alternate destination. In box 18, the code DEST/ appears, specifying the destination as "Estancia Olga" and its location as radial 170 from Córdoba VOR at 45 nautical miles.

In the final section, the fuel endurance is declared as five hours, four persons on board, one VHF radio installed, the aircraft's color and markings, and the name of the pilot in command.

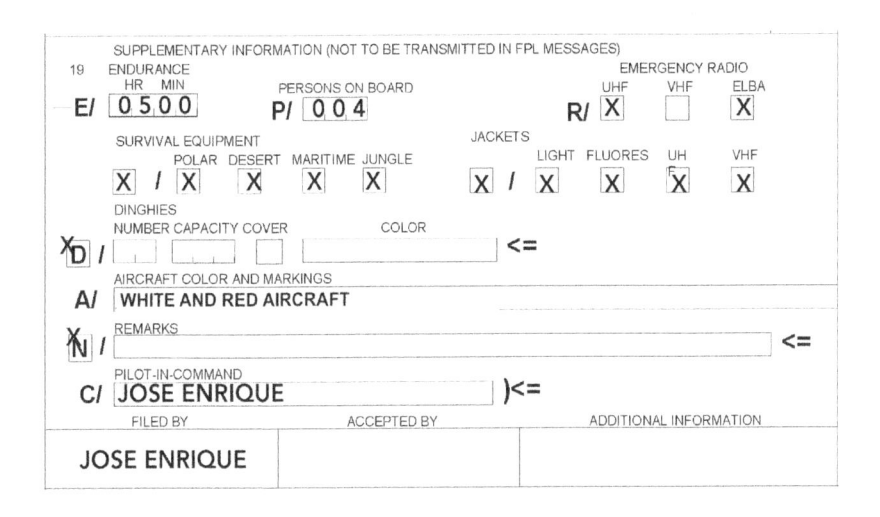

```
SUPPLEMENTARY INFORMATION (NOT TO BE TRANSMITTED IN FPL MESSAGES)
19  ENDURANCE                                    EMERGENCY RADIO
    HR  MIN                                      UHF    VHF    ELBA
E/  0 5 0 0      PERSONS ON BOARD    P/ 0 0 4    R/ X          X
    SURVIVAL EQUIPMENT                  JACKETS
    POLAR DESERT MARITIME JUNGLE        LIGHT FLUORES  UH   VHF
    X  /  X    X    X    X              X  /  X    X   X    X
    DINGHIES
    NUMBER CAPACITY COVER       COLOR
D/                                      <=
    AIRCRAFT COLOR AND MARKINGS
A/  WHITE AND RED AIRCRAFT
    REMARKS
N/                                                                <=
    PILOT-IN-COMMAND
C/  JOSE ENRIQUE                  )<=
    FILED BY          ACCEPTED BY         ADDITIONAL INFORMATION
    JOSE ENRIQUE
```

This concludes the flight plan for the first leg of the day. However, the pilot has submitted a second flight plan for the following day, when they will return from the estancia to Córdoba city.

The first section of the plan begins similarly to the previous form, declaring the aircraft registration, VFR flight rules, general type of flight, the same aircraft model, light turbulence category, standard equipment with transponder mode Charlie, and in the departure airport box, ZZZZ is declared since the estancia does not have an ICAO designator. The flight is scheduled to depart at 12:00 local time, maintaining the same speed and altitude as the previous day, and describes a direct route to Córdoba Airport.

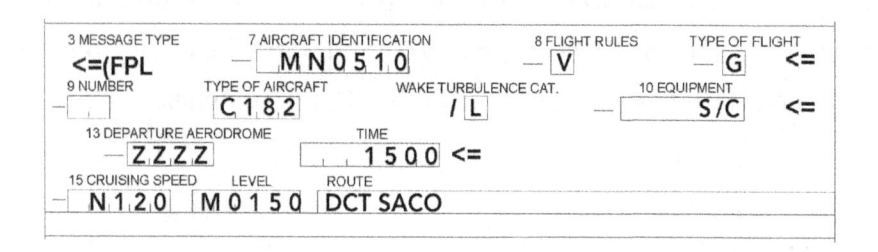

```
3 MESSAGE TYPE      7 AIRCRAFT IDENTIFICATION    8 FLIGHT RULES    TYPE OF FLIGHT
<=(FPL              — M N 0 5 1 0               — V              — G          <=
9 NUMBER            TYPE OF AIRCRAFT   WAKE TURBULENCE CAT.      10 EQUIPMENT
—                   C 1 8 2            / L                       — S /C      <=
13 DEPARTURE AERODROME             TIME
— Z Z Z Z                          1 5 0 0  <=
15 CRUISING SPEED   LEVEL      ROUTE
— N 1 2 0           M 0 1 5 0  DCT SACO
```

In the second section, the ICAO code of the destination airport is declared, with a flight time of 25 minutes and an alternative since it is a VFR flight. In this case, ZZZZ.

```
                              TOTAL EET
16 DESTINATION AERODROME      HR  MIN       ALTN AERODROME    2ND ALTN AERODROME
         S A C O              0 0 2 5            Z Z Z Z                         <=
18 OTHER INFORMATION
— DEPT/ ESTANCIA "OLGA" CBA170 45 NM
  ALTN/ ESTANCIA "OLGA" CBA170 45 NM
  DOF/ FLIGHT DATE SEPTEMBER 15
                                                                                 <=
```

In box 18, the codes DEPT/, ALTN/, and DOF/ appear. This clarifies that the departure aerodrome is the estancia "Olga" along with its location, and the alternative aerodrome is the same place.

Additionally, it is noted that this is a deferred flight plan, and the flight date will be the following day, September 15.

In the final section of supplementary information, a flight endurance of five hours is declared, with only one person on board, as the passengers were expected to stay at the field for a week. The aircraft is equipped with a single VHF radio, is white and red in color, and the pilot in command is the same individual who filed the flight plan.

```
     SUPPLEMENTARY INFORMATION (NOT TO BE TRANSMITTED IN FPL MESSAGES)
  19  ENDURANCE                                      EMERGENCY RADIO
        HR   MIN                                    UHF    VHF   ELBA
   E/ 0 5 0 0        PERSONS ON BOARD
                     P/ 0 0 1                   R/ |X|   | |    |X|
        SURVIVAL EQUIPMENT              JACKETS
           POLAR DESERT MARITIME JUNGLE          LIGHT FLUORES  UH   VHF
          X /  X       X        X              X /  X      X    X    X
        DINGHIES
        NUMBER CAPACITY COVER       COLOR
   XD/                                              <=
        AIRCRAFT COLOR AND MARKINGS
    A/  WHITE AND RED AIRCRAFT
        REMARKS
   XN/                                                              <=
        PILOT-IN-COMMAND
    C/  JOSE ENRIQUE                         )<=
          FILED BY              ACCEPTED BY           ADDITIONAL INFORMATION
        JOSE ENRIQUE
```

Basic Maneuvers

We define "Aerial Navigation" as the set of techniques that allow the pilot to take an aircraft from a starting point to a destination, having calculated all the variables involved.

104.- The "S-turns" maneuver over a road is initiated:

    A. At a point parallel to the reference.

    B. At a point perpendicular to the reference.

    C. At any point, considering the effect of the wind.

Scan this QR code with your smartphone or tablet to view the explanatory video of this maneuver on our YouTube channel.

105.- The "figure eights" maneuver over two points is initiated:

- A. Beside the two references.
- B. Between the two references.
- C. At any point, considering the effect of the wind.

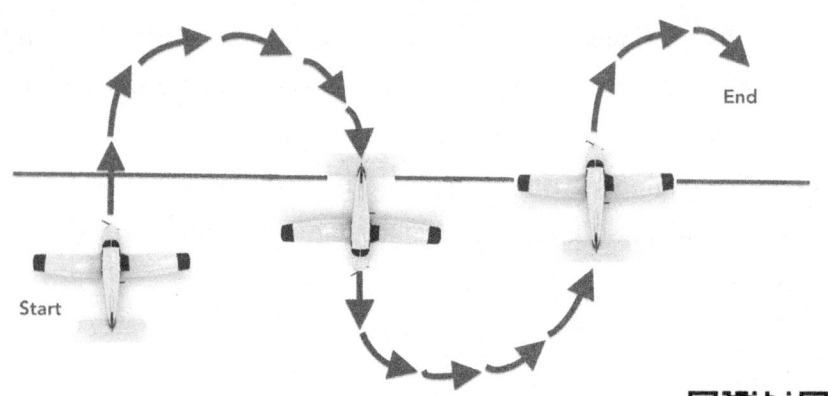

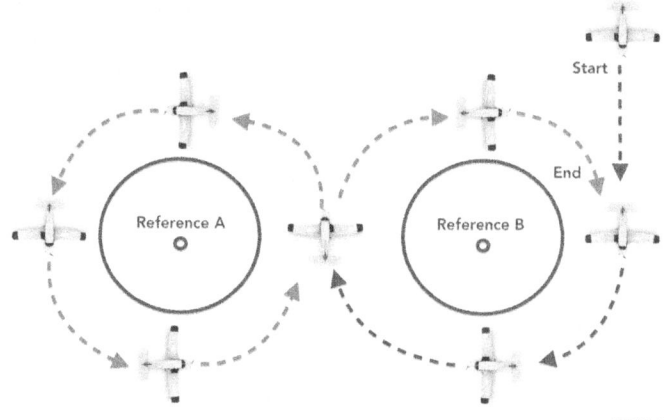

Scan this QR code with your smartphone or tablet to view the explanatory video of this maneuver on our YouTube channel.

106.- Drift correction is performed by:

   A. Modifying the aircraft's course.
   B. Modifying the flight path.
   C. Modifying the aircraft's heading.

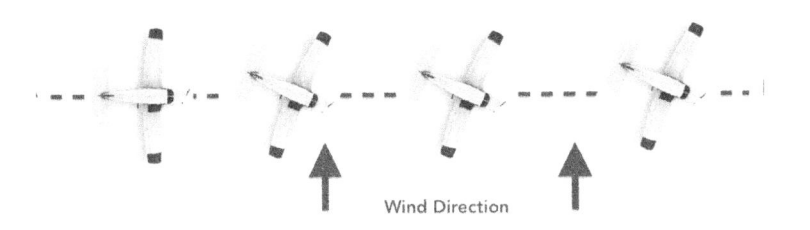

107.- A 90º approach begins:

   A. Perpendicular to the runway.
   B. Parallel to the runway.
   C. Either is correct, but always with the wind in favor.

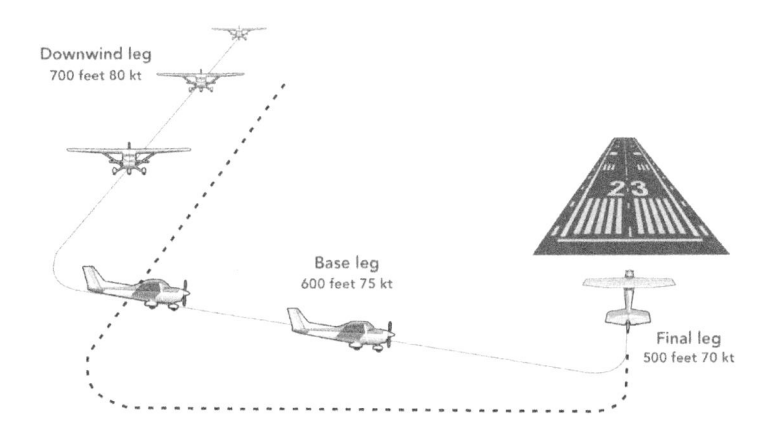

Escanea este código QR con tu Smartphone o Tablet para ver el video explicativo de esta maniobra en nuestro canal de YouTube.

108.- A 360° approach begins:

A. Perpendicular to the runway.
B. Parallel to the runway.
C. Either is correct, but always with the wind in favor.

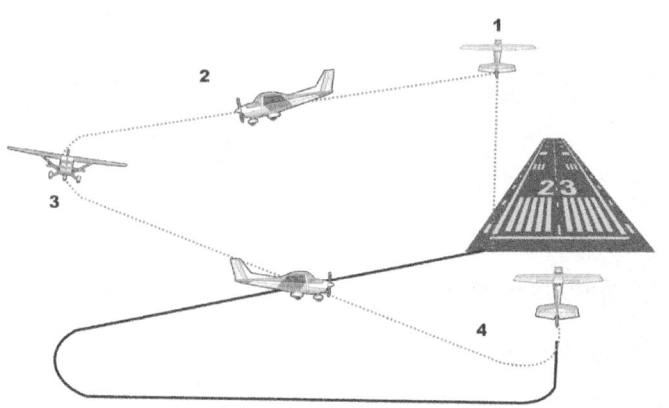

Escanea este código QR con tu Smartphone o Tablet para ver el video explicativo de esta maniobra en nuestro canal de YouTube.

109.- In a normal landing, power is reduced to minimum:

A. On final approach, with the runway in sight.
B. Once the aircraft is over the runway.
C. At any point in the traffic pattern.

Remember! This power reduction procedure may vary depending on wind direction and intensity during the final approach.

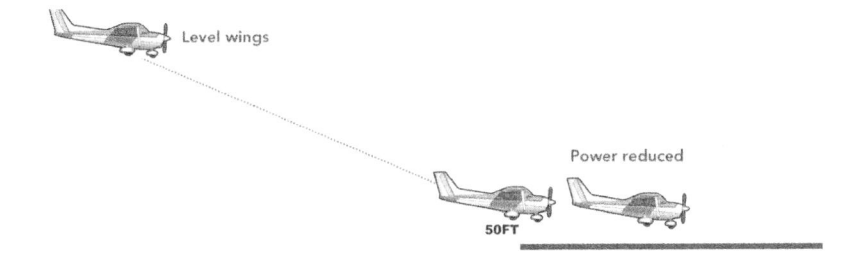

110.- To execute a missed approach procedure, the pilot must:

    A. Change the aircraft's attitude and climb.

    B. Change the aircraft's attitude, apply power, and climb.

    C. Apply power, change the aircraft's attitude, and climb.

Good job, Captain! You have successfully answered these 110 essential questions for your private pilot license. Remember that the pilot of an aircraft must possess extensive knowledge in various fields. These theoretical insights, which might seem unfamiliar today, will soon prove very useful. Don't forget, "knowledge doesn't take up space."

Correct Answers

01. **(B)**	23. **(A)**	45. **(C)**	67. **(B)**	89. **(B)**
02. **(C)**	24. **(B)**	46. **(C)**	68. **(C)**	90. **(C)**
03. **(A)**	25. **(A)**	47. **(B)**	69. **(A)**	91. **(B)**
04. **(B)**	26. **(B)**	48. **(B)**	70. **(C)**	92. **(B)**
05. **(B)**	27. **(A)**	49. **(C)**	71. **(C)**	93. **(B)**
06. **(A)**	28. **(C)**	50. **(C)**	72. **(B)**	94. **(B)**
07. **(C)**	29. **(C)**	51. **(C)**	73. **(B)**	95. **(A)**
08. **(B)**	30. **(B)**	52. **(C)**	74. **(C)**	96. **(B)**
09. **(C)**	31. **(B)**	53. **(B)**	75. **(B)**	97. **(A)**
10. **(B)**	32. **(C)**	54. **(B)**	76. **(B)**	98. **(A)**
11. **(B)**	33. **(C)**	55. **(C)**	77. **(C)**	99. **(C)**
12. **(B)**	34. **(A)**	56. **(A)**	78. **(A)**	100. **(A)**
13. **(B)**	35. **(B)**	57. **(B)**	79. **(B)**	101. -
14. **(A)**	36. **(A)**	58. **(C)**	80. **(B)**	102. -
15. **(C)**	37. **(B)**	59. **(C)**	81. **(B)**	103. -
16. **(C)**	38. **(B)**	60. **(B)**	82. **(B)**	104. **(B)**
17. **(A)**	39. **(B**	61. **(B)**	83. **(C)**	105. **(A)**
18. **(C)**	40. **(B)**	62. **(A)**	84. **(A)**	106. **(C)**
19. **(C)**	41. **(A)**	63. **(B)**	85. **(A)**	107. **(B)**
20. **(A)**	42. **(B)**	64. **(B)**	86. **(C)**	108. **(B)**
21. **(C)**	43. **(A)**	65. **(C)**	87. **(B)**	109. **(B)**
22. **(B)**	44. **(C)**	66. **(B)**	88. **(B)**	110. **(C)**

Chapter 2

In-Flight Knowledge for PPL

Private Pilot

Let's start your exam. We have arrived at the hangar, and it's time to prepare the aircraft for flight. The fuel tanks are full, we have conducted the briefing for the flight, and we have confirmed that the weather conditions will remain optimal. Now, all that remains is to conduct the aircraft preflight inspection, board the aircraft, and take off. To perform this task, I need to:

Option 1: Read the checklist, identify the items, put away the checklist, and begin the exterior inspection of the aircraft. Continue to page 78.

Option 2: Read the checklist and identify each item while performing the exterior inspection of the aircraft, always keeping the checklist in hand. Continue to page 85.

Option 3: Conduct the exterior inspection first since I have read the checklist during dozens of flights and already know each item. Additionally, the checklist was detailed during the briefing. Continue to page 82.

STOP!

Incorrect answer! This is your first mistake, but the exam has just begun. Remember that the maximum allowed is two mistakes, and with the third one, the exam will end. You will need to do your best on the remaining maneuvers to pass and obtain your license!

Taxiing the aircraft is a procedure that is often performed without following the manufacturer's instructions. The truth is that if we need to taxi our aircraft in extreme crosswind conditions, the effect would be the same as in flight; the wind could cause an undesirable displacement and push us off our intended path.

Although this is not our case now, imagine when you fly an A320, for example, which, due to its large size, would make the operation extremely complex. A slight deviation, and one of its wheels might go off the taxiway, causing damage to the landing gear.

Let this be a lesson! Now let's continue with your exam. Continue to page 84.

**A stumble is not a fall. Let's continue.
"You have 1 error."**

Knowing the correct bank angle for each turn is crucial for proper maneuvering. Let's climb to 3000 FT and evaluate stall maneuvers. We have reached 3000 FT, reduced power, gently raised the nose of the aircraft, and started decelerating. As we approach the stall speed, we recognize it and need to initiate the "Stall Recovery" maneuver by performing the following steps in the proposed order:

Option 1: Apply full power to regain speed and immediately reduce the angle of attack. Continue to page 83.

Option 2: Gently reduce the angle of attack immediately and apply progressive power to recover the lost speed and lift. Continue to page 142.

Option 3: Immediately lower the nose of the aircraft, recover speed, recover speed by negative angle of attack, and gently apply power while changing to a positive angle of attack to regain altitude. Continue to page 120.

STOP!

Incorrect answer! This is your first mistake, but the exam has just begun. Remember that the maximum allowed is two mistakes, and with the third one, the exam will end. You will need to do your best on the remaining maneuvers to pass and obtain your license!

External inspection procedures must be carried out with the checklist in hand. Today you are flying an aircraft with few systems, but when you fly a large commercial aircraft, there will be many items to check during your external inspection, and it would be easy to overlook something without the checklist in hand.

Keep in mind that a simple omission of a routine check, such as verifying the free movement of the ailerons (an item on the checklist), could lead to an accident if we take off without noticing that the ailerons are jammed and not functioning!

Your exam has just started, so don't panic! Let this be a lesson! Now let's continue with your exam. Continue to page 90.

Excellent Response!

Synchronizing the turn with the compass is essential to fly the correct headings! Well done, takeoff completed! We have reached 1000FT, and it's time to demonstrate some piloting skills. Let's perform 3 turns: one gentle, one normal, and one steep. For these maneuvers, we must:

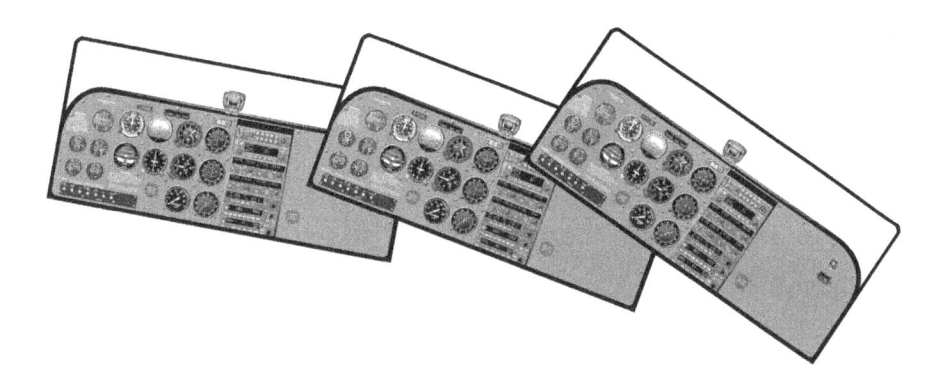

Option 1: Bank the aircraft to one side with 10°, 20°, and 30°, respectively, for each turn. Continue to page 87.

Option 2: Bank the aircraft to one side with 15°, 30°, and 45°, respectively, for each turn. Continue to page 99.

Option 3: Bank the aircraft to one side 10° or 15°, 20° or 30°, and 30° or 45°, depending on the wind intensity, to avoid undesired drift by adjusting the bank angle for each turn. Continue to page 91.

Let's continue calmly!
"You have 1 error."

During taxiing with a crosswind, you must maintain constant power and try to lower the wing on the windward side. Now, we are ready for takeoff. We position our aircraft on runway 23, and the "Directional Gyro" indicates heading 215°. Behind our plane, there is another aircraft at the holding point of runway 23, waiting to take off, so we need to take off as soon as possible! In this situation, the correct action is:

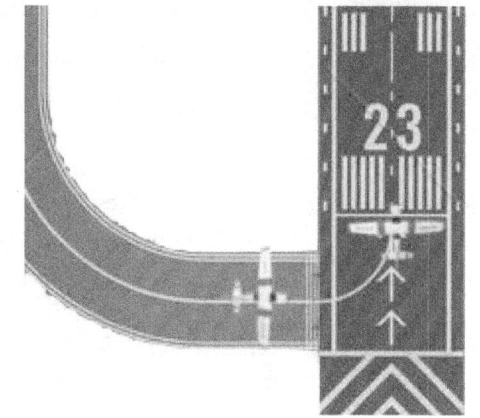

Option 1: Start the takeoff roll immediately to clear the runway. Continue to page 79.

Option 2: Take off and maintain the runway heading of 215° as indicated by the "Directional Gyro." Continue to page 96.

Option 3: Align the aircraft on the runway, calibrate the directional gyro to heading 230°, and begin the takeoff. Continue to page 93.

Well Done!

External inspection procedures should always be performed with the checklist in hand! Very well, we have started the engine of our aircraft and are ready to taxi. For this procedure, the wind is coming from the front, partially crosswind from the right. In a situation like this, we need to make a control input correction to avoid undesired drift during taxiing. In this case, we should:

Option 1: Neutral elevator and controls to the right. Continue to page 91.

Option 2: Neutral elevator and controls to the left. Continue to page 80.

Option 3: Neutral elevator and centered controls, but with a slight increase in power to counteract the force exerted by the wind. Continue to page 79.

You have 2 Errors!

Sorry! Your answer has been incorrect! This is your second error, and we haven't reached the sixth maneuver yet. You have reached the allowed limit. Turns around a point are very important to train your coordination skills between two parameters: maintaining a coordinated turn and keeping a constant altitude. Failing to develop this coordination ability could lead to an abnormal flight situation.

Remember, errors, no matter how small, can trigger an accident, causing irreversible damage. We will continue from here with two errors, but don't worry, if you perform all the remaining maneuvers correctly, your exam will be PASSED! So, keep your spirits up and move forward!

"Remember, you have 2 errors and it's the allowed limit."

This is your last chance. MAKE IT COUNT!

Return to page 116.

STOP!

Incorrect Answer! This is your first error, but the exam has just begun. Remember, the maximum allowed is two errors, and on the third one, we'll end the exam, so you'll need to strive hard on the remaining maneuvers to pass and obtain your license! Keep in mind that knowing the bank angle of a turn could save your life. The more we bank the aircraft, the more affected the lift becomes. Respecting the limits of each turn means adhering to a procedure designed to safeguard your life.

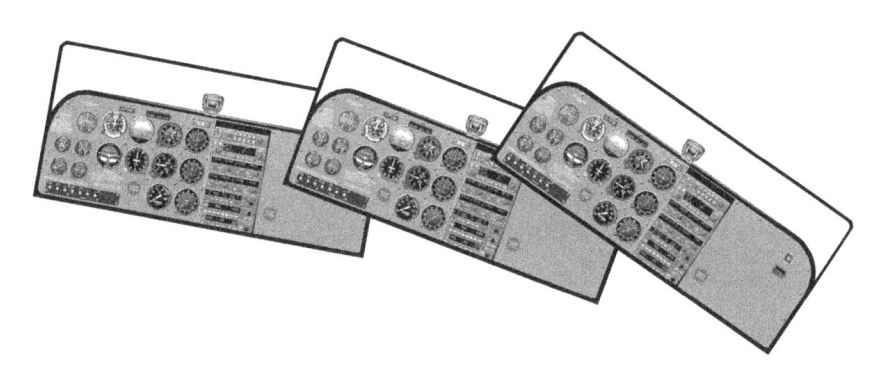

In certain commercial aircraft, there are bank angle protection systems, which allow the plane to bank only up to a certain angle, protecting flight safety. For now, we need to be aware of these limitations and respect them for safety.

Let this serve as a lesson! Now let's continue with your exam. Go to page 81.

STOP!

Incorrect Answer! This is your first error, but the exam has just begun. Remember, the maximum allowed is two errors, and on the third one, we'll end the exam, so you'll need to strive hard on the remaining maneuvers to pass and obtain your license!

Before initiating the takeoff roll, there are critical tasks such as completing the "Takeoff Briefing" before starting; observing the entire runway and climb path to confirm they are clear; conducting a final check of all instruments before taking the aircraft into the air, and several other tasks that might be added depending on the aircraft category. Prior to that, there is the holding point before entering the runway, and its purpose is precisely to "Wait" for the aircraft on the runway to complete all the aforementioned tasks, ensuring the safety of the operation.

Let this serve as a lesson! Now let's continue with your exam. Go to page 93.

STOP!

Incorrect Answer! This is your first error, but the exam has just begun. Remember, the maximum allowed is two errors, and on the third one, we'll end the exam, so you'll need to strive hard on the remaining maneuvers to pass and obtain your license. Understanding the purpose of a maneuver is often useful for understanding its execution. In the previous case, we attempted the maneuver of turns around a reference. This maneuver aims to practice coordinating turns at a constant speed and altitude without losing sight of the reference. Although it is an initial maneuver, it holds a certain complexity as it involves controlling multiple variables.

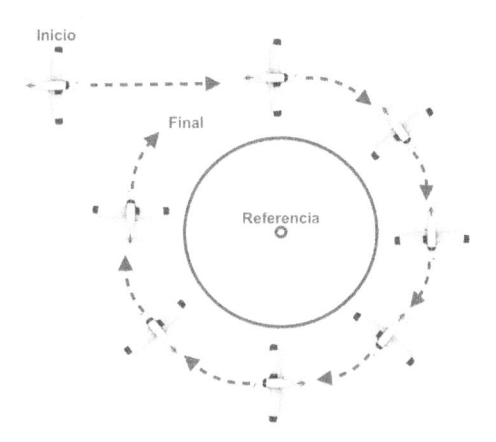

Keep in mind that this maneuver should always start with a tailwind! It is crucial that we learn to control the values of bank angle, altitude, and speed, all at the same time. Losing control of any of these values could cause the flight to become unstable and result in an undesirable or abnormal attitude.

Let this serve as a lesson! Now let's continue with your exam. Go to page 111.

Continue Forward

"You have 1 Error"

Don't be alarmed as this is just the beginning! Remember that exterior inspection procedures should always be carried out with the checklist in hand. Let's start the engine and begin taxiing. In this situation, the wind is coming from the front, partially crosswind from the right. To handle this situation, we need to make the following control corrections to prevent undesired displacement during taxiing. In this case, we should:

Option 1: Neutral elevator trim and right control inputs. (See page 84)

Option 2: Neutral elevator trim and left control inputs. (See page 94)

Option 3: Neutral elevator trim and centered controls, but with a slight increase in power to counteract the wind's force. (See page 79)

Great Job!

Excellent response! You have two correct answers and no errors. During crosswind taxiing, you should maintain constant power and try to lower the wing on the windward side. Now we are ready for takeoff. We have positioned our aircraft on Runway 23, and the "Heading Indicator" shows a heading of 215°. Behind our aircraft, there is another aircraft waiting for takeoff, so we need to get airborne as soon as possible! In this situation, the correct action is:

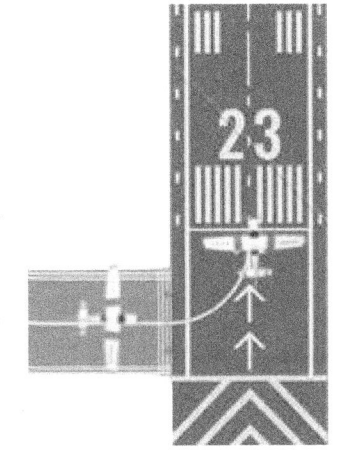

Option 1: Immediately begin the takeoff roll to clear the runway. (See page 85)

Option 2: Take off and maintain the runway heading of 215° as indicated by the "Heading Indicator". (See page 88)

Option 3: Align the aircraft on the runway, adjust the heading indicator to 230°, and then start the takeoff. (See page 83)

STOP!

Incorrect answer! This is your first error, but the exam has just begun. Remember that the maximum allowed is two errors, and on the third error, the exam will end. You will need to focus on the remaining maneuvers to pass and obtain your license! Practicing **"Stall Recovery"** is crucial. As pilots, we must not only recognize a stall situation but also know how to recover from such an extreme condition.

Remember that an aircraft "Enters a Stall" due to an EXCESSIVE ANGLE OF ATTACK! This decreases the speed and leads to a loss of lift. Following the previous order, the first action to take in a stall situation is to reduce the angle of attack by lowering the nose of the aircraft. This will increase the speed, and finally, apply power to regain the original flight path!

Let's use this as a lesson! Now, continue with your exam. Proceed to page 120.

Let's continue!
"You have 1 Error"

Synchronizing the turn with the compass is crucial for flying the correct headings! Now that we've reached 1000 feet, it's time to demonstrate some piloting skills. Let's perform 3 turns: one shallow, one standard, and one steep. For these maneuvers, we should:

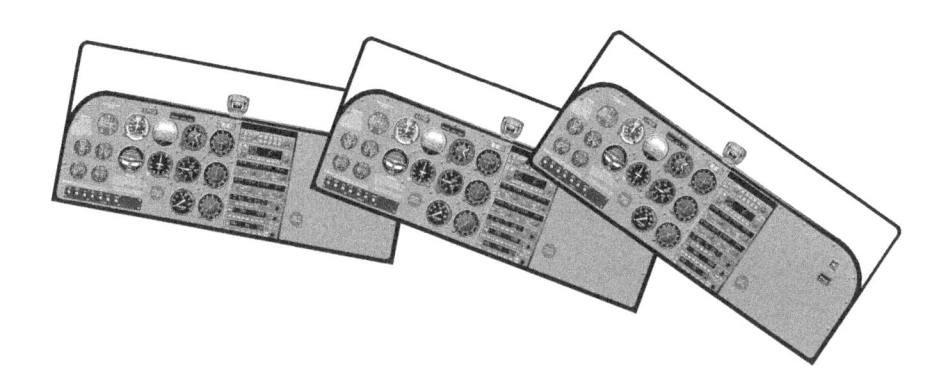

Option 1: Tilt the aircraft to one side with 10°, 20°, and 30° respectively for each turn. See page 103.

Option 2: Tilt the aircraft to one side with 15°, 30°, and 45° respectively for each turn. See page 81.

Option 3: Tilt the aircraft to one side with 10° or 15°, 20° or 30°, and 30° or 45° depending on the intensity of the wind, to avoid unwanted displacement by adjusting the tilt for each turn. See page 84.

STOP!
You have 2 Errors!

I'm sorry! Your answer was incorrect! This is your second error, and we haven't reached the third maneuver yet. You have reached the allowed limit of two errors.

Es Understanding crosswind taxiing operations is crucial as it can impact flight safety. Although it may not seem apparent now, maintaining high standards is meant to help you improve and become proficient.

Keep in mind that even small mistakes can lead to an accident, causing irreversible damage. We will continue from here with two errors, but don't be alarmed; if you perform all the remaining maneuvers correctly, your exam will be PASSED! So, stay encouraged and keep going!

"Remember, you have 2 errors and this is the allowed limit"

It's your last chance. MAKE THE MOST OF IT! Go back to page 98.

Well done! Keep it up!

An appropriate drift correction helps keep your flight path on track. Let's continue with the flight and climb back up to 3000 FT. One of the most serious situations we can face is, undoubtedly, an engine failure.

Let's simulate an engine failure. For this, I will reduce all power. Ready? Here we go!

"Ok, power reduced."
What should we do now?

Option 1: First, establish the glide speed with wings level. Then, look for the most suitable place for an emergency landing. See page 101.

Option 2: First, establish the glide speed with wings level. Then, look for the nearest place to make an emergency landing. See page 109.

Option 3: First, establish the glide speed and gently initiate the descent. Then, look for the most suitable and closest place to make an emergency landing. See page 119.

**STOP!
You have 2 Errors!**

Do not be alarmed; this is just the beginning! It is crucial to understand the correct takeoff procedure. The runway orientation indicates the heading the aircraft should follow upon takeoff. If the heading indicated by the compass is different, we will be flying with incorrect information, which could lead to disorientation.

Keep in mind that even small mistakes can lead to an accident, causing irreversible damage. We will continue from here with two errors, but don't be alarmed; if you perform all the remaining maneuvers correctly, your exam will be PASSED! So, stay encouraged and keep going!

"Remember, you have 2 errors and this is the limit allowed"

It's your last chance. MAKE THE MOST OF IT!

Return to page 100

KEEP going!
"You have 2 Errors"

Proper drift correction ensures that your flight path remains on course. Let's continue with the flight. We will climb back to 3000 FT. One of the most critical situations we may face is, undoubtedly, an engine failure. Let's simulate an engine emergency. I will reduce all power. Ready? Here we go!

"Ok, power reduced."
What should we do now?

Option 1: First, establish glide speed with wings level. Then, look for the most suitable place to make an emergency landing, attempting a restart. Refer to page 116.

Option 2: First, establish glide speed with wings level. Then, look for the nearest place to make an emergency landing, attempting a restart. Refer to page 117.

Option 3: Establish glide speed and begin descent. Then, look for the most suitable and closest place to make an emergency landing, attempting a restart. Refer to page 133.

Continue calmly!
"You have 2 Errors"

Crosswind taxiing should maintain constant power and attempt to lower the wing on the windward side. Now, we are ready for takeoff. We have positioned our aircraft on runway 23, and the "Heading Indicator" shows 215°. Behind our aircraft, another aircraft is waiting at the runway 23 holding point, ready for takeoff, so we need to depart as soon as possible! In this situation, the correct action is:

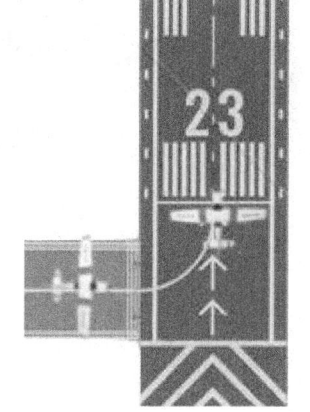

Option 1: Immediately start the takeoff roll to clear the runway. Refer to page 117.

Option 2: Take off and maintain the runway heading of 215° as indicated by the "Heading Indicator." Refer to page 90.

Option 3: Align the aircraft on the runway, calibrate the heading indicator to 230°, and then initiate takeoff. Refer to page 100.

Excellent!

Knowing the bank angle for each turn is crucial for performing correct turns. Let's ascend to 3000 feet and evaluate the "Stall" maneuvers. Having reached 3000 feet, reduce power, gently raise the aircraft's nose, and begin decelerating. Once you reach the stall speed, recognize it, and initiate the "Stall Recovery" maneuver by following the steps in the proposed order:

Option 1: Apply maximum power to recover speed and immediately reduce the angle of attack. Refer to page 91.

Option 2: Gently reduce the angle of attack first and then progressively apply power to recover from the loss of speed and lift. Refer to page 92.

Option 3: Immediately lower the nose of the aircraft, recover speed by reducing the angle of attack, and then gently apply power while smoothly transitioning to a positive angle of attack to regain altitude. Refer to page 109.

Perfect! Let's continue

"You have 2 Errors"

You've reached 1000 feet, and it's time to demonstrate your piloting skills. We need to perform three turns: one gentle, one normal, and one steep. For these maneuvers, we should:

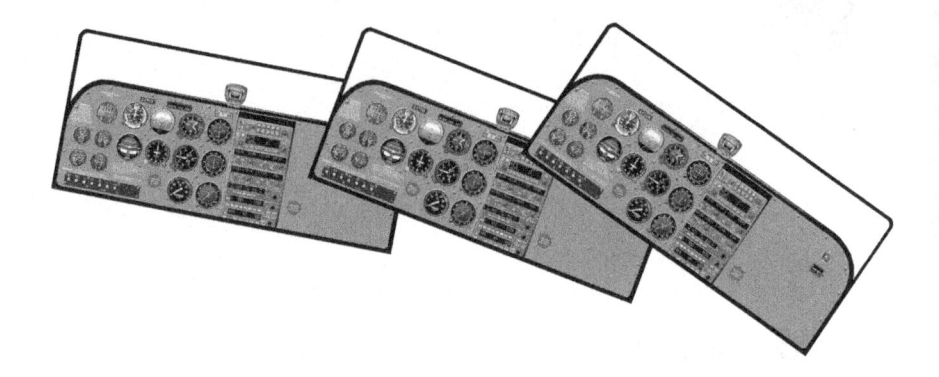

Option 1: Bank the aircraft 10°, 20°, and 30°, respectively, for each turn. Refer to page 98.

Option 2: Bank the aircraft 15°, 30°, and 45°, respectively, for each turn. Refer to page 110.

Option 3: Bank the aircraft 10° or 15°, 20° or 30°, and 30° or 45°, depending on wind intensity, to avoid unwanted displacement by adjusting the bank angle for each turn. Refer to page 117.

STOP!

Incorrect response! This is your first error, but the exam has just begun. Remember, the maximum allowed is two errors, and upon reaching three, the exam will end. Therefore, you need to perform well in the remaining maneuvers to pass and obtain your license.

Here's something interesting about engine failure emergencies: they are not as severe as many believe! Remember, an aircraft flies due to speed, not power. If the engine fails and stops, it's inevitable that speed will decrease, but we will have a few moments to calmly consider the next steps. The key is not to immediately decide to descend but to do so only when necessary to regain speed. The manufacturer specifies a glide speed, which is appropriate for these situations. If we are above this speed at the time of failure, we can maintain altitude until speed reaches the glide speed. This gives us a few extra seconds to assess the situation, attempt to restart the engine, find a suitable landing area, or simply calm our nerves!

Let this be a lesson! Now, let's continue with your exam. Please turn to page 122.

Great Job!

The "Slip" maneuver allows us to descend in a controlled manner at a higher rate while maintaining a constant speed. Now, demonstrate how to perform a slip. We will execute this maneuver and descend to 1500 FT. Afterward, we will return to the runway to practice some approaches. We will perform the slip as follows:

Option 1: Bank the wing towards the side where the wind is coming from. Apply opposite rudder and control the descent at a constant speed. Continue on page 95.

Option 2: Bank the wing to one side, regardless of the wind direction. Apply opposite rudder and aim for the maximum descent at a constant speed. Continue on page 132.

Option 3: Bank the wing towards the side from which the wind is coming. Apply opposite rudder and control the descent at a constant speed. Continue on page 108.

STOP!
You have 2 Errors!

I'm sorry! Your response was incorrect! This is your second error and we haven't reached the fifth maneuver yet. You have reached the allowed limit. Mastering coordinated turns is essential for good aircraft handling. We need to understand the bank angle for each turn to respect our aircraft's limitations. Exceeding the maximum bank angle of a turn could cause the aircraft to lose lift and begin to fall!

Keep in mind that even small mistakes can lead to an accident, causing irreversible damage. We will continue from here with two errors, but don't be alarmed; if you perform all the remaining maneuvers correctly, your exam will be PASSED! So, stay encouraged and keep going!

"Remember, you have 2 errors and this is the limit allowed"
It's your last chance. MAKE THE MOST OF IT!

Return to page 110

Let's continue!
"You have 1 Error"

Well done. We have reached the final approach for runway 04. This stage requires the highest precision due to the close proximity to the terrain during the final descent, so BE CAREFUL. We continue the descent, crossing 200 feet, and the airspeed is under control, but upon looking up, we see that runway 04 is considerably displaced to our left even though we are still flying on heading 040°. The wind effect has played a trick on us, so we need to:

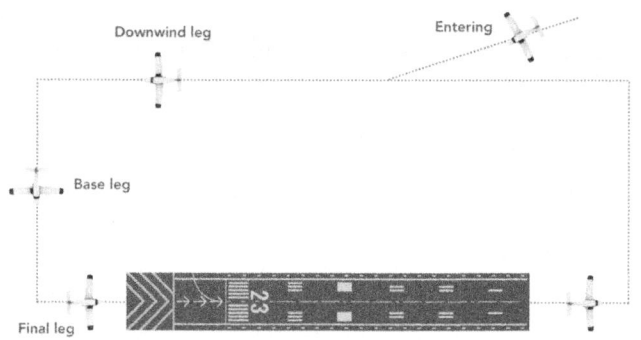

Option 1: Maintain heading 040° and execute a missed approach. Refer to page 129

Option 2: Abandon heading 040° and adopt a new heading to align with the runway for landing, while maintaining airspeed. Refer to page 125

Option 3: Stop the descent until the runway is directly ahead, correcting the heading, then continue with the descent and landing on runway 04. Refer to page 107

Excellent!

Excellent response! You have six correct answers and no errors. After completing the previous maneuver, we continue flying straight and level with a forward reference, trying to maintain constant altitude and airspeed. After a few seconds, we notice that our reference is no longer directly ahead but slightly displaced to the left:

The wind effect is causing an unwanted drift. To correct this, we should:

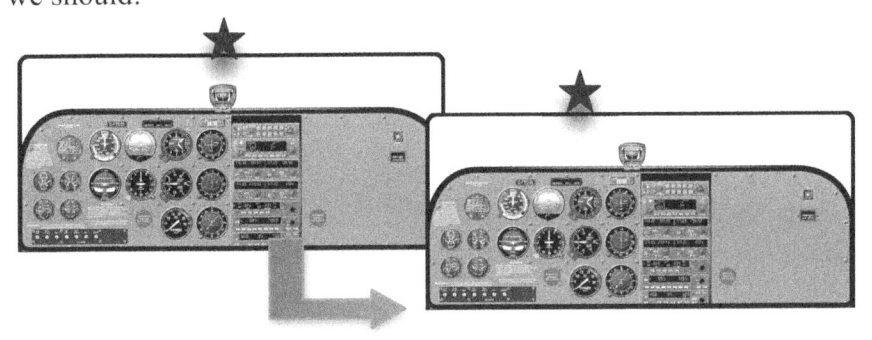

Option 1: Apply right pedal pressure to correct the drift caused by the wind. Refer to page 130

Option 2: Correct the heading to the left enough to keep the reference directly ahead. Refer to page 95

Option 3: Maintain the current heading, increasing power to overcome the wind force causing the unwanted drift. Refer to page 99

Great Job!

Well done! Excellent decision. In any situation that jeopardizes the landing, we must perform a go-around. Now that the decision is made, let's execute this maneuver like the professionals!

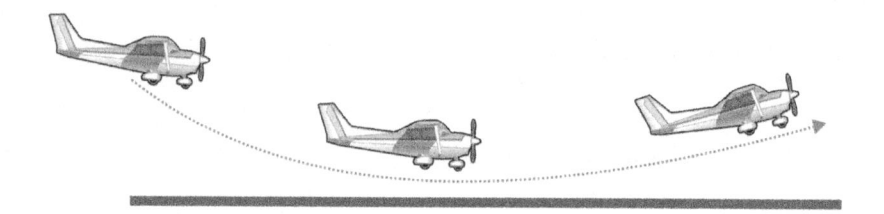

Option 1: First, retract the flaps gradually, apply full power, and initiate an immediate climb. Refer to page 108

Option 2: Stop the descent. Apply full power and initiate the climb. Once a safe altitude is reached, retract the flaps. Refer to page 150

Option 3: Apply full power. Initiate an immediate climb with the wings level and retract the flaps gradually. Refer to page 147

Complex maneuver!
"You have 1 error"

The "slip" maneuver allows us to descend in a controlled manner and at a higher rate without increasing speed. Very well, demonstrate how we should perform a slip. Let's execute this maneuver and descend to 1500 feet. Afterwards, we will return to the runway to practice some approaches. We will perform the slip as follows:

Option 1: Bank the aircraft towards the side where the wind is coming from. Apply opposite rudder and control the descent at a constant speed. Refer to page 122.

Option 2: Bank the aircraft to one side, regardless of the wind direction. Apply opposite rudder and seek the maximum descent rate at a constant speed. Refer to page 145.

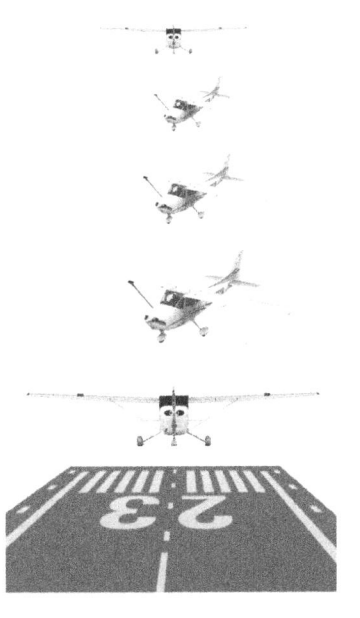

Option 3: Bank the aircraft towards the side from which the wind is coming. Apply opposite rudder and control the descent at a constant speed. Refer to page 112.

Excellent work!

Let's head back to the runway and practice some approaches. Remember, we must adhere to the traffic pattern for each approach. Descend to 500 feet. We have the runway directly ahead, perpendicular to our flight path. We need to enter the initial leg of runway 04 for a standard 90° approach.

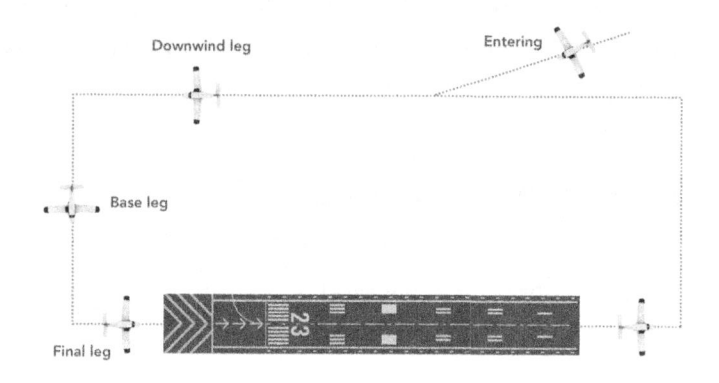

Option 1: Reduce speed to the manufacturer's recommended value and enter the initial leg at any sector of it, flying heading 220°. Refer to page 136.

Option 2: Reduce speed and join the initial leg after passing abeam runway 22, flying heading 220°, maintaining altitude throughout the leg. Refer to page 121.

Option 3: Reduce speed and join the initial leg after passing abeam runway 04, flying heading 220° and seeking an appropriate descent for the initial leg as indicated by the local traffic pattern. Refer to page 119.

Great Job!

Once the stall has been recovered, let's proceed with the next maneuver. Choose a reference point and perform "Turns Around a Point." In our flight area, the wind is blowing strongly from the north. We are currently flying heading 220° and you have already selected a reference point for the maneuver. The next step will be:

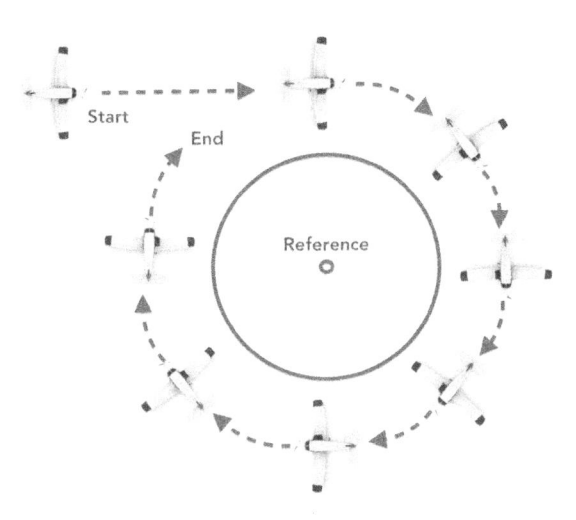

Option 1: Maintain heading until abeam the chosen reference, maintaining altitude, and then initiate the maneuver. See page 105.

Option 2: Turn to heading 360° and, when abeam the reference, initiate the maneuver. See page 89.

Option 3: Turn to heading 180° and, when abeam the reference, initiate the maneuver. See page 83.

**A stumble is not a fall. Let's continue.
"You have 2 Errors."**

Understanding the bank angle for each turn is crucial for performing turns correctly. Let's climb to 3000 feet and evaluate the "Stall" maneuvers. Alright, we've reached 3000 feet, reduce power, gently raise the nose of the aircraft, and begin decelerating. Upon reaching the stall speed, recognize it and initiate the "Stall Recovery" maneuver by performing the following steps in the proposed order:

Option 1: Apply full power to regain speed and immediately reduce the angle of attack. See page 100.

Option 2: Gradually reduce the angle of attack and progressively apply power to recover from the loss of speed and lift. See page 117.

Option 3: Lower the nose of the aircraft immediately, regain speed by reducing the angle of attack, and smoothly apply power while transitioning to a positive angle of attack to recover altitude. See page 114.

If you need a break, just let me know!
"You have 1 error"

Remember that the previous maneuver starts with a tailwind. Once the previous maneuver is completed, continue flying straight and level with a reference ahead, trying to maintain altitude and constant speed. After a few seconds, we notice that our reference is no longer ahead but slightly off to the left:

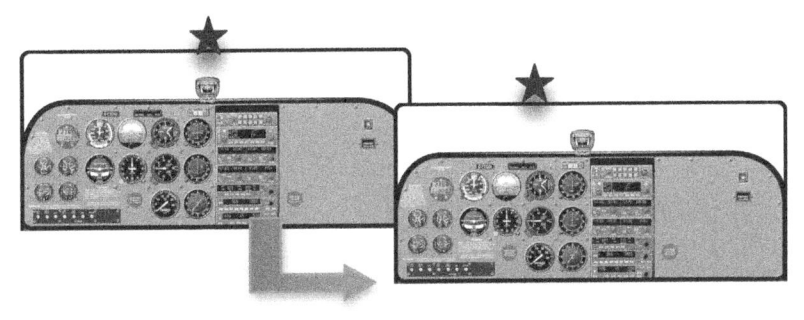

The wind is causing an unwanted drift. To correct this, we should:

Option 1: Apply right rudder pressure to correct the drift caused by the wind. See page 120.

Option 2: Adjust the heading to the left enough to keep the reference directly ahead. See page 118.

Option 3: Maintain the current heading, increasing power to overcome the wind's force that causes the unwanted drift. See page 124.

Let's continue!

"You have 1 Error"

Excellent response! Let's return to the runway and practice some approaches. Remember, we must respect the traffic pattern in each approach. Descend to 500 FT. The runway is directly ahead, perpendicular to our flight path. We need to join the initial leg for runway 04 to perform a traditional 90° approach. To do this:

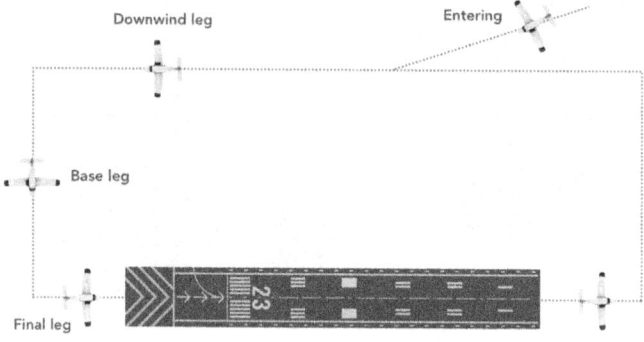

Downwind leg

Entering

Base leg

Final leg

Option 1: Reduce speed to the manufacturer's recommended setting and enter the initial leg from any point on it while flying heading 220°. See page 141.

Option 2: Reduce speed and join the initial leg after passing abeam runway 22, flying heading 220°, without changing altitude throughout the leg. See page 104.

Option 3: Reduce speed and join the initial leg after passing abeam runway 04, flying heading 220°, and seek an appropriate descent for the initial leg as indicated by the local traffic pattern. See page 122.

Complex maneuver!
"You have 2 Errors"

The "Slip" maneuver allows us to descend in a controlled manner and at a higher rate without increasing speed. Let's perform this maneuver and descend to 1500 FT. Afterward, we will return to the runway to practice some approaches. We will conduct the slip as follows:

Option 1: Bank the wing towards the direction from which the wind is blowing. Apply opposite rudder and control the descent at a constant speed. See page 133.

Option 2: Bank the wing to one side, regardless of the wind direction. Apply opposite rudder and seek the maximum descent at a constant speed. See page 117.

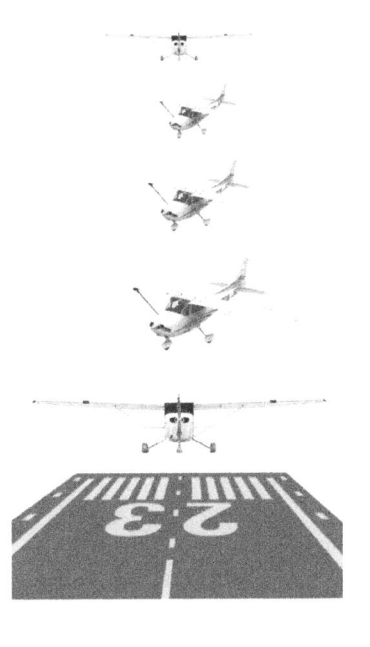

Option 3: Bank the wing towards the direction from which the wind is coming. Apply opposite rudder and control the descent at a constant speed. See page 135.

Let's continue!

"You have 2 Errors"

It is crucial to learn how to recognize a "stall" and how to recover from it, as your life will depend on it! Let's proceed by choosing a reference point and performing "Turns Around a Point." In our flight area, the wind is strong and coming from the north. We are currently flying on a heading of 220°, and you have chosen a reference for the maneuver. The next step will be:

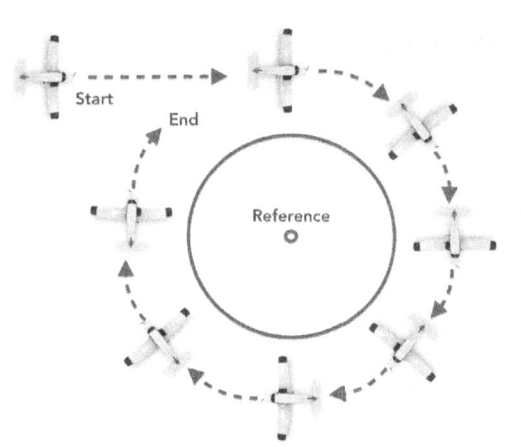

Option 1: Maintain the heading until passing abeam the chosen reference, keeping altitude and then initiate the maneuver. See page 116.

Option 2: Turn to heading 360° and initiate the maneuver when abeam the reference. See page 100.

Option 3: Turn to heading 180° and initiate the maneuver when abeam the reference. See page 117.

STOP!

Incorrect Answer! This is your first error, but the exam has just begun. Remember, the maximum allowed is two errors, and at the third error, the exam will end. You will need to put extra effort into the remaining maneuvers to pass and obtain your license. Remember, we were imagining a situation with inoperative "Flaps"!

Any aircraft, regardless of its size, can descend rapidly by using an excessive negative angle of attack (nosing down), but this would inevitably increase speed. In other words, no aircraft can descend and decelerate simultaneously. This situation was improved by implementing lift augmentation devices like Flaps, which allow for an increased negative angle of attack without gaining speed, but only up to a certain point.

In our case, the correct answer is to perform a side-slip maneuver, as it allows for controlled descent without a significant increase in speed. However, like the use of Flaps, it also has its limitations!

Let this be a lesson! Now, let's continue with your exam. Proceed to page 107.

If you need a break, just let me know!
"You have 2 Errors"

Remember that the previous maneuver started with a tailwind. Once the previous maneuver is completed, continue flying straight and level with a reference in front, maintaining altitude and speed. After a few seconds, you notice that the reference is no longer directly ahead but slightly to the left: The wind is causing undesired drift. To correct this, you should:

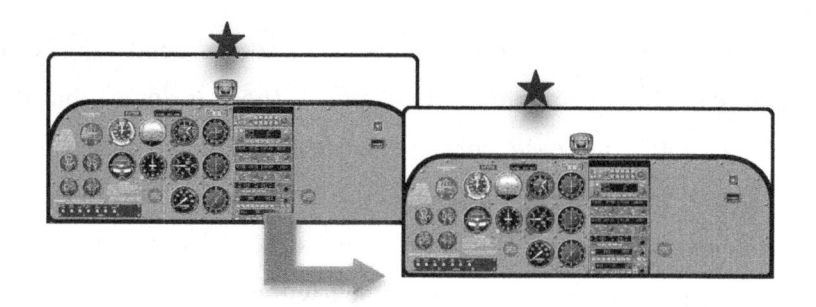

Option 1: Apply right rudder pressure to correct the drift caused by the wind. Refer to page 110

Option 2: Adjust the heading to the left enough to realign the reference in front. Refer to page 97

Option 3: Maintain the current heading and increase power to counteract the wind's drift. Refer to page 117

Incorrect Answer!

I'm sorry! Your response was incorrect, and this is your third accumulated error. Unfortunately, your exam is:

FAILED!

As humans, we will make mistakes throughout our lives, and the important thing is not to try to eliminate errors completely but to learn how to address them when they occur, and rest assured they will! Now, don't be alarmed, try again, and you will surely pass successfully. As mentioned at the beginning, the maximum number of allowed errors is two, but it's not a random number; it's designed to make us pause and evaluate if we need to learn a bit more or not.

At this point, you should return to the start of the book to begin your exam again. I suggest reviewing the previous incorrect answers and learning from them.

GOOD LUCK!
CAPT. THOMAS

END OF FLIGHT. Continue on page 79.

Let's continue!

"You have 1 Error"

Proper drift correction ensures that your flight path remains on course. Let's continue with the flight, climb back to 3000 feet. One of the most critical situations we may encounter is an engine failure. Let's simulate an engine emergency. I will reduce all power now. Ready? Here we go!"

Understood. Power reduced.
What will we do now?

Option 1: First, establish glide speed with wings level. Then, find the most suitable emergency landing site, attempting a restart. See page 120.

Option 2: First, establish glide speed with wings level. Then, find the nearest emergency landing site, attempting a restart. See page 128.

Option 3: Establish glide speed and initiate descent. Then, find the most suitable and closest emergency landing site, attempting a restart. See page 122.

Great Job!

In an engine failure emergency, the most important thing is to establish the manufacturer-recommended glide speed and then find the most suitable place for an emergency landing. Without realizing it, we have come very close to the runway and are still at 3000 feet. Imagine that the flaps are stuck and cannot be operated. We need to descend to 1500 feet as quickly as possible to avoid moving further away from the runway. Be cautious with the speed; it must be kept reduced.

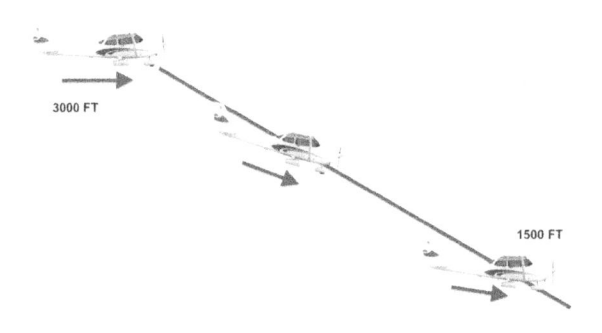

Option 1: First, reduce the power, then lower the nose and aim for a controlled descent. See page 105.

Option 2: First, reduce the power to minimum, then establish a continuous descent rate to maintain speed. See page 115.

Option 3: First, reduce the power and then perform a controlled slip to manage the speed. See page 102.

Let's continue!
"You have 1 Error"

It is crucial to learn to recognize a stall and know how to recover from it, as your life will depend on it! Let's continue by choosing a reference and performing "Turns Around a Point." In our flight area, the wind is blowing strongly from the north. We are currently flying on a heading of 220° and you have chosen a reference for the maneuver. The next step will be:

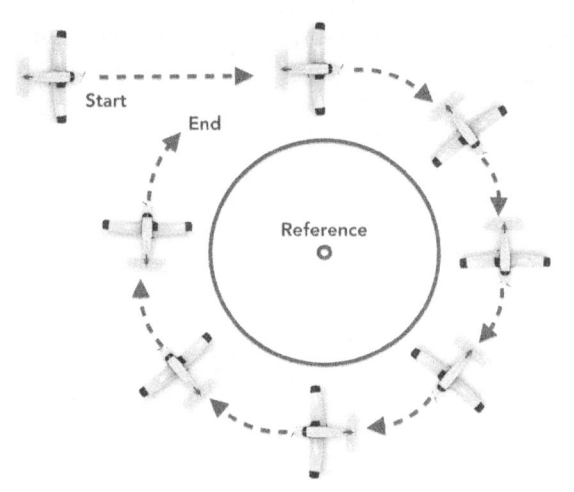

Option 1: Maintain heading until passing abeam the chosen reference, keeping altitude, and then initiate the maneuver. See page 111.

Option 2: Turn to heading 360° and, when flying abeam the reference, initiate the maneuver. See page 86.

Option 3: Turn to heading 180° and, when flying abeam the reference, initiate the maneuver. See page 93.

Excellent work!

Well done. We have reached the FINAL segment of runway 04. This phase requires the highest precision due to the proximity to the ground during the final descent, so BE CAREFUL. We continue the descent, passing 200 FT, observe the airspeed and it remains under control, but when we look up, we see that runway 04 is quite displaced to our left even though we are still flying on heading 040°. The wind effect has played a trick on us, so we must:

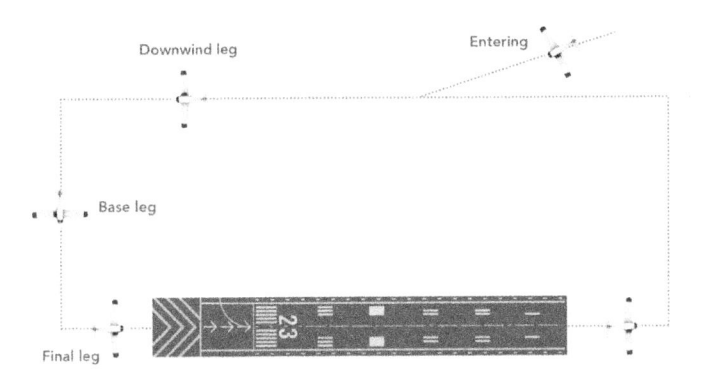

Option 1: Maintain heading 040° and execute a go-around. See page 106.

Option 2: Abandon heading 040° and adopt a new heading to align with the runway to continue the landing, while maintaining airspeed. See page 102.

Option 3: Stop the descent until the runway is directly ahead by correcting the heading, then continue the descent and landing on runway 04. See page 126.

Complex maneuver!
"You have 1 error"

In an engine failure emergency, the most important thing is to achieve the glide speed specified by the manufacturer and then locate the most SUITABLE place for an emergency landing. Unintentionally, we have approached the runway closely while still flying at 3000 FT. Assume that the FLAPS are stuck and cannot be operated. We need to descend to 1500 FT as quickly as possible to avoid moving further away from the runway. Be cautious with the speed; WE MUST keep it reduced.

Option 1: First, reduce power, then lower the nose and seek a controlled descent. See page 111.

Option 2: First, reduce power to minimum, then adopt a continuous descent rate to maintain speed. See page 140.

Option 3: First, reduce power and then perform a slip while controlling the speed. See page 107.

Perfect! Let's continue
"You have 2 Errors"

We have reached the FINAL approach for Runway 04. This stage requires the highest precision due to the proximity to the ground during the final descent, so BE CAREFUL. We continue the descent, passing 200 FT, and observe that the speed is under control. However, upon looking up, we see that Runway 04 is significantly displaced to our left, even though we are still flying on a heading of 040°. The wind effect has caused an issue. Therefore, we should:

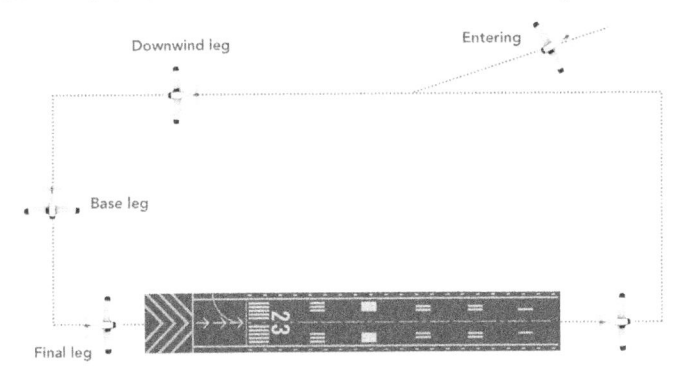

Option 1: Maintain heading 040° and execute a go-around. See page 134.

Option 2: Deviate from heading 040° and adopt a new heading that will lead us to the runway to continue the landing, while maintaining speed. See page 117.

Option 3: Stop the descent until the runway is directly ahead by correcting the heading, then continue the descent and landing on Runway 04. See page 135.

STOP!
You have 2 Errors!

I'm sorry! Your answer was incorrect! This is your second error, and we haven't even reached the sixth maneuver yet. You've reached the allowed limit. Correcting drift is crucial for visual navigation. Remember, the wind causes a displacement effect on our flight path, and if not corrected in time, we would not reach the intended location. This could affect fuel consumption and create an emergency situation.

Keep in mind that even small mistakes can lead to an accident, causing irreversible damage. We will continue from here with two errors, but don't be alarmed; if you perform all the remaining maneuvers correctly, your exam will be PASSED! So, stay encouraged and keep going!

"Remember, you have 2 errors and this is the limit allowed"
It's your last chance. MAKE THE MOST OF IT!

Return to page 97

STOP!
You have 2 Errors!

I'm sorry! Your answer was incorrect! This is your second error, and we haven't even reached the sixth maneuver yet. You have reached the allowed limit. The final segment of the approach is the highest risk phase; we are flying at low altitude and minimal speed. This is where a VITAL maneuver, the "Go-Around" or missed approach, becomes crucial. Remember this every time, in the final segment, if the parameters are not correct.

Keep in mind that even small mistakes can lead to an accident, causing irreversible damage. We will continue from here with two errors, but don't be alarmed; if you perform all the remaining maneuvers correctly, your exam will be PASSED! So, stay encouraged and keep going!

"Remember, you have 2 errors and this is the limit allowed"
It's your last chance. MAKE THE MOST OF IT!
Return to page 134

STOP!

Incorrect answer! This is your first error, but the exam has just begun. Remember, the maximum allowed is two errors, and we will end the exam at the third error, so you will need to be diligent with the remaining maneuvers to pass and obtain your license.

Don't forget that in any risky situation, especially at low altitude, we must take actions that ensure safety. In our case, the correct procedure is to perform a go-around or missed approach. Remember that in the final segment of the approach, we must have the runway ahead, maintain controlled speed, and keep the descent angle within the established parameters for the maneuver. Any deviation from these parameters indicates that the approach has destabilized, and a go-around should be performed to retry the approach or consider alternative options.

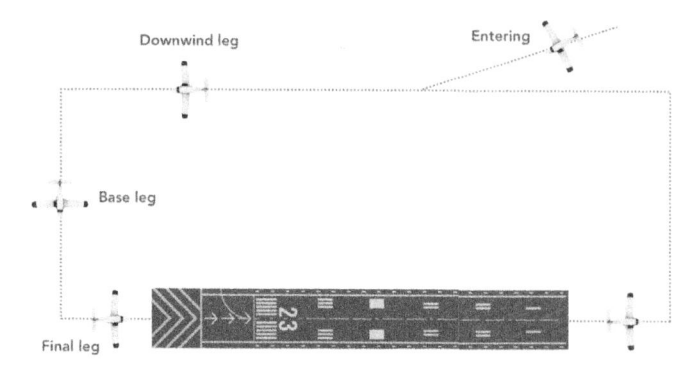

Anyway, let this be a lesson! Now, let's continue with your exam. Proceed to page 129.

Let's continue!
"You have 1 Error"

Let's rejoin the traffic pattern to perform a 180° approach. Remember that the elevation of our airfield is 100FT. Keep this information in mind for the maneuver. We will land on runway 04. Show me how to execute a 180° approach.

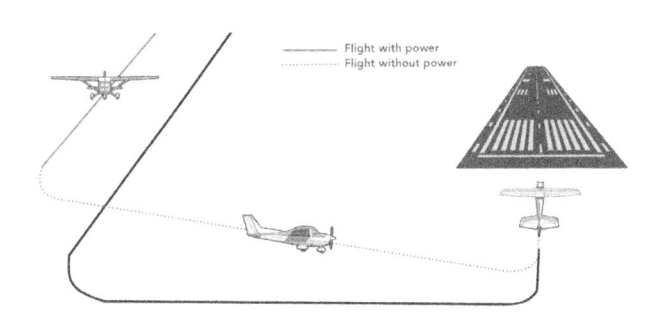

Option 1: Begin the maneuver on the initial leg at 600FT altitude. When abeam the start of runway 04, reduce power and initiate a glide, performing a 180° turn to align with the final approach leg. See page 146.

Option 2: Begin the maneuver on the initial leg at 600FT altitude. After flying abeam the start of runway 04, reduce power and initiate a glide to the base leg, and then to the final approach leg. See page 139.

Option 3: Begin the maneuver abeam the start of runway 04 on a heading of 040° at 600FT altitude. Reduce power and initiate a glide to the base leg, and then to the final approach leg. See page 104.

Let's continue

"You have 2 Errors"

Sorry! Your answer was incorrect! This is your second error and we haven't reached the sixth maneuver yet. You have reached the allowed limit. Emergency procedures are the most critical maneuvers in aviation. Remember a basic concept: the aircraft flies by airspeed, so if the engine fails, you will have some time to choose a suitable emergency landing area. It will be essential to manage your airspeed.

Keep in mind that even small mistakes can lead to an accident, causing irreversible damage. We will continue from here with two errors, but don't be alarmed; if you perform all the remaining maneuvers correctly, your exam will be PASSED! So, stay encouraged and keep going!

"Remember, you have 2 errors and this is the limit allowed"
It's your last chance. MAKE THE MOST OF IT!
Return to page 133

Perfect! Let's continue
"You have 1 Error"

In any situation that jeopardizes the landing, we must perform a go-around. Now, the decision is made! Let's discard the possibility of landing and execute this maneuver. Show me how professionals do it!

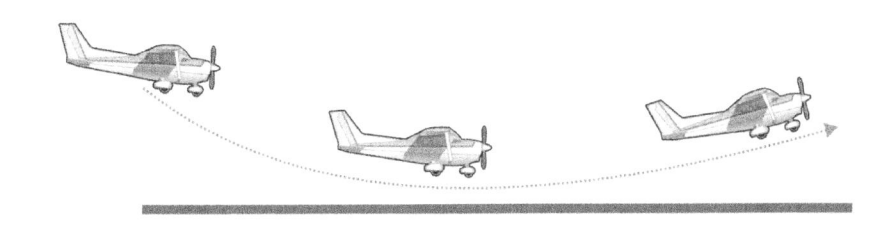

Option 1: Initially, retract the flaps gradually, apply full power, and immediately initiate the climb. Refer to page 137.

Option 2: Stop the descent. Apply full power and initiate the climb. Once a safe altitude is reached, retract the flaps. Refer to page 127.

Option 3: Apply full power. Immediately initiate a climb with the wings level and retract the flaps gradually. Refer to page 112.

STOP!

Incorrect Answer! This is your first error, and the exam has just begun. Remember, the maximum allowed is two errors, and at the third, we will conclude, so you will need to focus on the remaining maneuvers to pass and obtain your license. You are taking your first steps in aviation, and drift correction is an extremely important point. You might hear that drift correction is often done by pressing the pedal on the side where the wind is blowing to exert more force in that direction, but this is not entirely correct, as if you have to navigate 200 NM with a crosswind, you could not maintain the pedal pressed for that long!

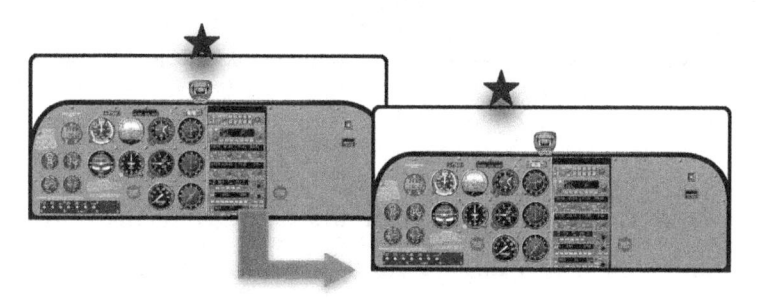

The proper way to correct drift caused by the wind is to adjust the heading by a certain number of degrees to counteract the unwanted displacement! It's a very common mistake, so don't be alarmed, but remember that in all cases, corrections will always be made towards the side from which the wind is blowing!

Let this be a lesson! Now let's continue with your exam. Proceed to page 118.

Good Job!
"You have 2 Errors"

Let's rejoin the traffic pattern to perform a 180° approach. Remember that the elevation of our airfield is 100 feet. Keep this information in mind for the maneuver. Let's land on Runway 04. Show me how to execute a 180° approach.

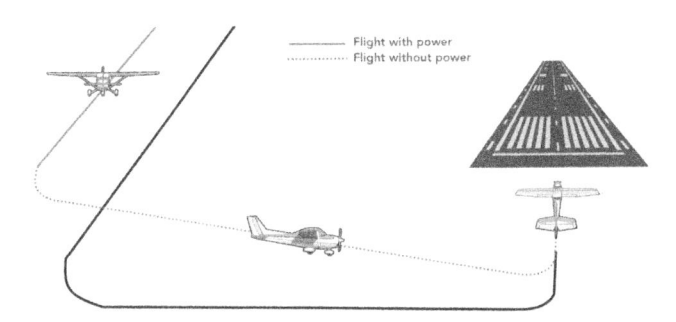

Option 1: Initiate the maneuver on the initial leg at 600 feet altitude. Upon passing abeam the start of Runway 04, reduce power and initiate a glide while making a 180° turn to join the final approach. Refer to page 117.

Option 2: Initiate the maneuver on the initial leg at 600 feet height. After flying abeam the start of Runway 04, reduce power and begin a glide to the base leg, then proceed to the final approach. Refer to page 167.

Option 3: Initiate the maneuver abeam the start of Runway 04 with heading 040° at 600 feet height. Reduce power and begin a glide to the base leg, then proceed to the final approach. Refer to page 134.

STOP!

Incorrect Answer! This is your first error, but the exam has just begun. Remember, the maximum allowed is two errors, and at the third, we will finish, so you will need to put in extra effort for the remaining maneuvers to pass and obtain your license.

A side-slip is a maneuver requiring caution, as it involves descending in an abnormal flight attitude, but in a controlled manner.

By applying the control stick to one side and pressing the opposite rudder pedal, we are creating a "side-slip" while in a descent attitude. This prevents the speed from increasing rapidly, but be careful, as excessive side-slipping could actually increase speed and lead to an uncontrolled situation.

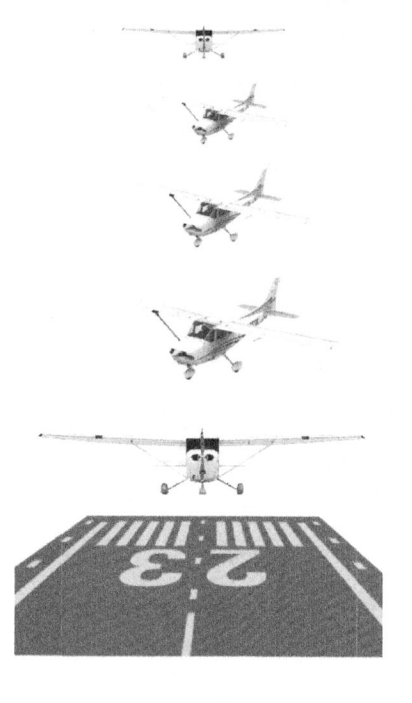

Let this serve as a lesson! Now, let's continue with your exam. Proceed to page 112.

Good Job!
"You have 2 Errors"

In an engine failure emergency, the most important thing is to establish the glide speed recommended by the manufacturer and then find the most suitable place for an emergency landing! Without realizing it, we have approached the runway closely and are still at 3000 FT. Let's assume that the flaps are jammed and cannot be operated. We need to descend to 1500 FT as quickly as possible to avoid drifting too far from the runway. Be careful with the speed; we must keep it reduced.

Option 1: First, reduce the power, then lower the nose and seek a controlled descent. Refer to page 114.

Option 2: First, reduce the power to the minimum, then adopt a continuous descent rate to maintain speed. Refer to page 117.

Option 3: First, reduce the power and then perform a side-slip while controlling the speed. Refer to page 113.

Perfect! Let's continue
"You have 1 Error"

In any situation that jeopardizes the landing, we must execute a go-around. Now that the decision is made, let's proceed with the go-around maneuver. Show me how professionals do it!

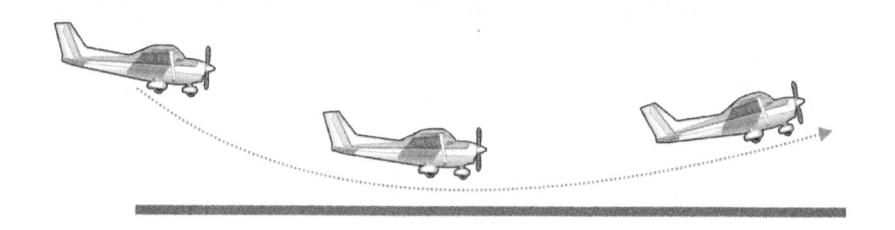

Option 1: Initially, gradually retract the flaps, apply full power, and begin the climb immediately. Refer to page 123.

Option 2: Stop the descent. Apply full power and initiate the climb. Once at a safe altitude, retract the flaps. Refer to page 131.

Option 3: Apply full power. Start an immediate climb with wings level and gradually retract the flaps. Refer to page 117.

Great!
"You have 2 Errors"

Excellent response! Let's return to the runway and practice some approaches. Remember that we must adhere to the traffic pattern on each approach. Let's descend to 500FT. We have the runway directly ahead, perpendicular to our flight path. We need to join the base leg for runway 04 to perform a standard 90° approach. Here's how:

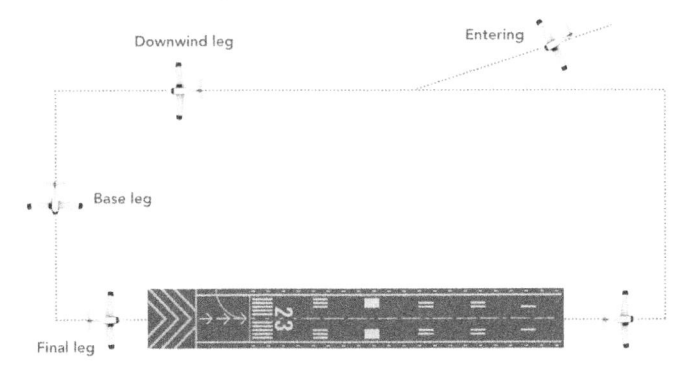

Option 1: Reduce speed to the manufacturer's recommended value and enter the base leg at any point, flying on a heading of 220°. Continue on page 117.

Option 2: Reduce speed and join the base leg when abeam runway 22, maintaining a heading of 220° without changing altitude throughout the leg. Continue on page 123.

Option 3: Reduce speed and join the base leg when abeam runway 04, flying on a heading of 220° and adjusting descent appropriately for the base leg as per the local traffic pattern. Continue on page 113.

STOP!

Incorrect response! This is your first error, but the exam has just begun. Remember that the maximum allowed is two errors, and at the third, we will stop, so you'll need to perform well in the remaining maneuvers to pass and obtain your license.

Remember that the purpose of the traffic pattern is to organize the various aircraft operating at the aerodrome before landing, ensuring traffic control to maximize safety. Therefore, it is crucial to respect all segments of the pattern. The pattern begins with the upwind leg, where each aircraft must join parallel to the runway, opposite to the landing direction. Keep in mind that the minimum legal altitude for this leg is 500FT. Next is the base leg, where we continue the descent perpendicular to the landing path, followed by the final leg, with the runway ahead, at approach speed and configuration as established by the manufacturer in the operating manuals.

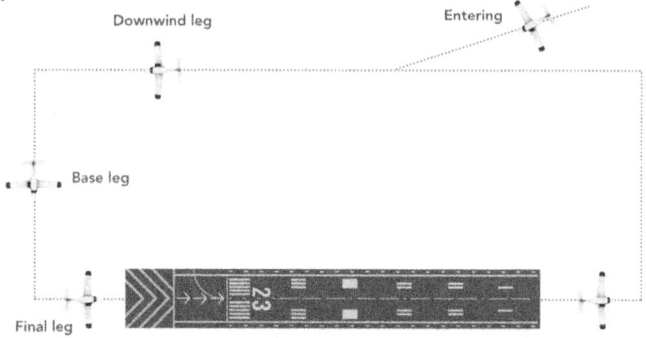

In any case, let this be a lesson! Now, let's continue with your exam. Continue on page 104.

STOP!
You have 2 Errors!

Sorry! Your response has been incorrect! This is your second error, and we haven't even reached the sixth maneuver yet. You've reached the allowed limit.

Always remember that the go-around maneuver is designed to elevate and distance you from the ground as quickly as possible. This maneuver should always be considered during the final approach and in any situation you deem risky. When in doubt or uncertainty, GO AROUND!

Keep in mind that even small mistakes can lead to an accident, causing irreversible damage. We will continue from here with two errors, but don't be alarmed; if you perform all the remaining maneuvers correctly, your exam will be PASSED! So, stay encouraged and keep going!

"Remember that you have 2 errors, and that is the allowed limit." This is your last chance. MAKE THE MOST OF IT!

Go back to page 131.

STOP!

the third, we will stop, so you'll need to perform well in the remaining maneuvers to pass and obtain your license.

The purpose of practicing this maneuver is to develop the necessary skills to land the aircraft during the approach while maintaining a full glide. To do this, when flying abeam the runway numbers, we must reduce power and continue to land without its assistance.

Over time, we will develop a sense of distance and altitude based on our aircraft's gliding capability, which will be crucial. This sense is vital because, in an emergency, it could be extremely useful. Never forget that you must know your aircraft as well as you know yourself to recognize its limitations and respect them!

In any case, let this be a lesson! Now, let's continue with your exam.

Continue on page 139.

Good Job!
"You have 1 Error"

We have landed on runway 04 after executing a successful 180° approach. We will vacate the runway, take the first taxiway, and return to the holding point for runway 04 for another takeoff. We enter runway 04, align the aircraft on a heading of 040°, and after applying full power, begin the takeoff roll. However, during the roll, the airspeed indicator stops functioning, and we decide to perform an RTO (rejected takeoff). To do this, we should:

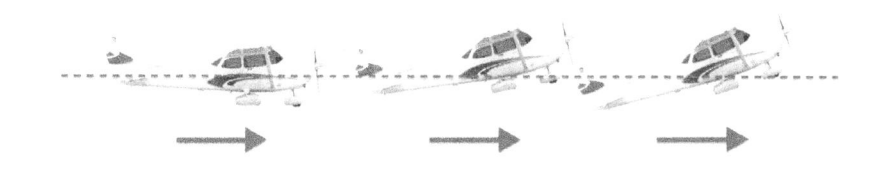

Option 1: Reduce power, maintain the roll along the centerline, and bring the aircraft to a complete stop. Vacate the runway to assess the situation. Continue on page 149.

Option 2: Decelerate the aircraft to a suitable taxi speed, return to the apron, and assess the situation. Continue on page 143.

Option 3: If there is no remaining runway, continue the takeoff normally, with the exception that we will enter the traffic pattern to land immediately in order to assess the cause of the failure. Continue on page 129.

STOP!
You have 2 Errors!

Sorry! Your response has been incorrect! This is your second error, and we haven't even reached the sixth maneuver yet. You've reached the allowed limit. In aircraft without flaps, the sideslip maneuver allows for a steeper descent rate than usual. This is a maneuver that requires great caution because we will be operating the aircraft in an abnormal but controlled flight attitude. Be careful with this maneuver!

Keep in mind that even small mistakes can lead to an accident, causing irreversible damage. We will continue from here with two errors, but don't be alarmed; if you perform all the remaining maneuvers correctly, your exam will be PASSED! So, stay encouraged and keep going!

"Remember that you have 2 errors, and that is the allowed limit." This is your last chance. MAKE THE MOST OF IT!

Go back to page 113.

STOP!
You have 2 Errors!

Sorry! Your response has been incorrect! This is your second error, and we haven't even reached the sixth maneuver yet. You've reached the allowed limit. Keep in mind that the purpose of the traffic pattern is to ensure the safety of the aircraft operating within it. We must know its segments and directions perfectly so that all aircraft follow the same procedures, avoiding mid-air collisions, which could result in a serious accident.

Keep in mind that even small mistakes can lead to an accident, causing irreversible damage. We will continue from here with two errors, but don't be alarmed; if you perform all the remaining maneuvers correctly, your exam will be PASSED! So, stay encouraged and keep going!

"Remember that you have 2 errors, and that is the allowed limit." This is your last chance. MAKE THE MOST OF IT!

Go back to page 123.

STOP!
You have 2 Errors!

Sorry! Your response has been incorrect! This is your second error, and we haven't even reached the sixth maneuver yet. You've reached the allowed limit. Stall recovery is one of the most important maneuvers in your career, as it could save your life. It is ESSENTIAL that you master this technique, as errors in stall recovery can cause the aircraft to "break up" in seconds and spiral out of control.

Keep in mind that even small mistakes can lead to an accident, causing irreversible damage. We will continue from here with two errors, but don't be alarmed; if you perform all the remaining maneuvers correctly, your exam will be PASSED! So, stay encouraged and keep going!

"Remember that you have 2 errors, and that is the allowed limit." This is your last chance. MAKE THE MOST OF IT!

Go back to page 114.

STOP!
You have 2 Errors!

Sorry! Your response has been incorrect! This is your second error, and we haven't even reached the sixth maneuver yet. You've reached the allowed limit. Keep in mind that it is better to resolve a problem on the ground than in the air. Based on this concept, the RTO is a maneuver that can prevent an in-flight incident by detecting an anomaly during the takeoff roll. Don't forget, better on the ground than in the air!

Keep in mind that even small mistakes can lead to an accident, causing irreversible damage. We will continue from here with two errors, but don't be alarmed; if you perform all the remaining maneuvers correctly, your exam will be PASSED! So, stay encouraged and keep going!

"Remember that you have 2 errors, and that is the allowed limit." This is your last chance. MAKE THE MOST OF IT!

Go back to page 157.

**STOP!
You have 2 Errors!**

Sorry! Your response has been incorrect! This is your second error, and we haven't even reached the sixth maneuver yet. You've reached the allowed limit. Short field takeoffs help the aircraft lift its wheels off the ground as quickly as possible. Failing to follow this procedure on a contaminated runway could result in the aircraft not becoming airborne in time, leading to a potential risk, incident, or accident.

Keep in mind that even small mistakes can lead to an accident, causing irreversible damage. We will continue from here with two errors, but don't be alarmed; if you perform all the remaining maneuvers correctly, your exam will be PASSED! So, stay encouraged and keep going!

"Remember that you have 2 errors, and that is the allowed limit." This is your last chance. MAKE THE MOST OF IT!

Go back to page 156.

STOP!
You have 2 Errors!

Sorry! Your response has been incorrect! This is your second error, and we haven't even reached the sixth maneuver yet. You've reached the allowed limit. In aircraft without flaps, the sideslip maneuver allows for a steeper descent rate than usual. This is a maneuver that requires great caution because we will be operating the aircraft in an abnormal but controlled flight attitude. Be very careful with this maneuver!

Keep in mind that even small mistakes can lead to an accident, causing irreversible damage. We will continue from here with two errors, but don't be alarmed; if you perform all the remaining maneuvers correctly, your exam will be PASSED! So, stay encouraged and keep going!

"Remember that you have 2 errors, and that is the allowed limit." This is your last chance. MAKE THE MOST OF IT!

Go back to page 135.

STOP!
You have 2 Errors!

Sorry! Your response has been incorrect! This is your second error, and we haven't even reached the sixth maneuver yet. You've reached the allowed limit. The 180° approach aims to train your skills in flying a glide, controlling the parameters, and performing a coordinated turn to enter the final leg. It simulates an emergency in the pattern, so it is crucial that you master this technique perfectly.

Keep in mind that even small mistakes can lead to an accident, causing irreversible damage. We will continue from here with two errors, but don't be alarmed; if you perform all the remaining maneuvers correctly, your exam will be PASSED! So, stay encouraged and keep going!

"Remember that you have 2 errors, and that is the allowed limit." This is your last chance. MAKE THE MOST OF IT!

Go back to page 167.

Good job!

Excellent response! Let's rejoin the traffic pattern to perform a 180° approach. Remember that the elevation of our airfield is 100FT. Take this into account for the maneuver. Let's land on runway 04. Show me how to execute a 180° approach.

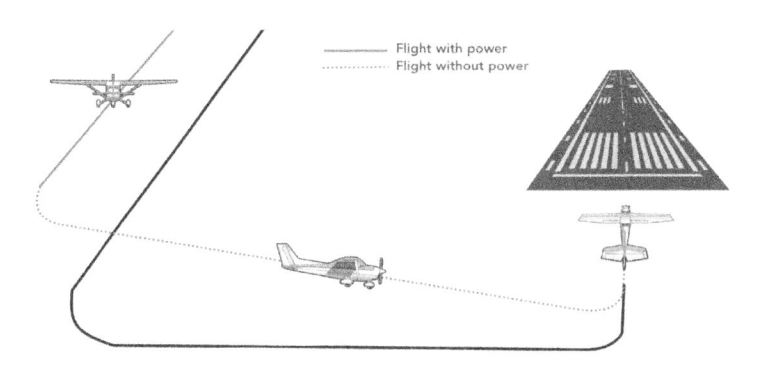

Flight with power
Flight without power

Option 1: Begin the maneuver on the upwind leg at 600FT altitude. When abeam the start of runway 04, reduce power and initiate a glide, making a 180° turn to join the final leg. Continue on page 138.

Option 2: Begin the maneuver on the upwind leg at 600FT altitude. After flying abeam the start of runway 04, reduce power and initiate a glide to the base leg and then to the final leg. Continue on page 170.

Option 3: Begin the maneuver abeam the start of runway 04 on a heading of 040° at 600FT altitude. Reduce power and initiate a glide to the base leg and then to the final leg. Continue on page 121.

STOP!

Incorrect response! This is your first error, but the exam has just begun. Remember that the maximum allowed is two errors, and at the third, we will stop, so you'll need to perform well in the remaining maneuvers to pass and obtain your license.

Don't be alarmed; we will continue with the takeoff and your exam. However, we need to remember that short field takeoff procedures aim to achieve minimum lift speed over the shortest distance.

To achieve this, we use flaps, which extend the wing surface and, as a result, increase lift or achieve it at lower speeds. Remember that the lift formula is: $L = CL \times 1/2\rho \times SUP \times V^2$. As aircraft operators, we can influence the last two values in this formula: wing surface area (using flaps) and speed (attitude and power).

In any case, let this be a lesson! Now, let's continue with your exam.

Continue on page 153.

Good Job!
"You have 1 Error"

The RTO maneuver is of extreme importance for the safety of everyone. After clearing runway 04, we will taxi to a different runway and take off from runway 13.

Be careful! This runway is in very poor condition, so we need to perform a short field takeoff. Do you remember this maneuver? Let's review its execution together. We need to extend the flaps. Keeping the brakes applied, we apply full power, release the brakes, and initiate the takeoff roll. My question is: this maneuver is performed with flaps extended in all cases; do you remember why?

Option 1: Extending the flaps increases lift and allows the aircraft to become airborne over a shorter distance during the takeoff roll. Continue on page 144.

Option 2: Extending the flaps delays the stall speed, allowing the aircraft to sustain at a lower speed and become airborne sooner with a shorter takeoff distance than usual. Continue on page 153.

Option 3: Extending the flaps increases the wing surface area, which enhances lift and allows the aircraft to take off at a lower speed, resulting in a shorter takeoff distance. Continue on page 139.

STOP!

Incorrect response! This is your first error, but the exam has just begun. Remember that the maximum allowed is two errors, and at the third, we will stop, so you'll need to perform well in the remaining maneuvers to pass and obtain your license.

Let's imagine the worst-case scenario. Suppose we are at 100FT altitude and, for some reason, we need to perform a go-around. The first thing we need to achieve is full engine power to climb as quickly as possible. Once we have gained separation from the ground, we can consider other actions, such as initiating a turn, beginning flap retraction, among other tasks, but the priority is always to apply full power, keep the wings level, and gain altitude as soon as possible.

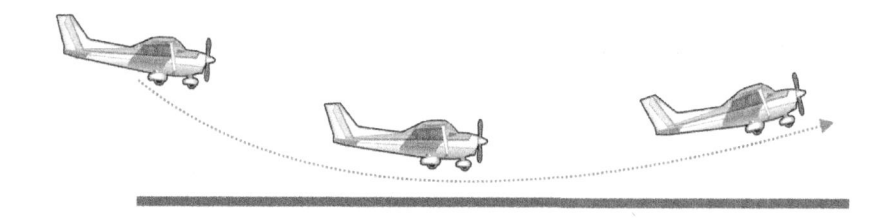

Let this be a lesson! Now, let's continue with your exam.

Continue on page 127.

STOP!

Incorrect response! This is your first error, but the exam has just begun. Remember that the maximum allowed is two errors, and at the third, we will stop, so you'll need to perform well in the remaining maneuvers to pass and obtain your license.

5000 FT

3000 FT

2000 FT

There are theoretical concepts that might lead to errors but generally won't escalate into more serious issues. However, some theoretical concepts are of vital importance as they directly affect the safety of operations, not just yours but for everyone. This is one of them. Remember that the famous "Transition Layer" is an imaginary layer at 1000FT where the altimeter setting must be adjusted accordingly. If climbing, you switch from QNH to QNE, and if descending, you switch from QNE to QNH. This ensures that all aircraft operate with the same altimeter setting.

Let this be a lesson! Now, let's continue with your exam.

Continue on page 164.

STOP!
You have 2 Errors!

Sorry! Your response has been incorrect! This is your second error, and we haven't even reached the sixth maneuver yet. You've reached the allowed limit. Remember that the transition layer is a 1000FT airspace intended for adjusting altimeter settings, transitioning from flight levels to altitudes and vice versa. You must have this concept clear; otherwise, you would be flying with incorrect altimeter information, which could lead to a collision in flight.

Keep in mind that even small mistakes can lead to an accident, causing irreversible damage. We will continue from here with two errors, but don't be alarmed; if you perform all the remaining maneuvers correctly, your exam will be PASSED! So, stay encouraged and keep going!

"Remember that you have 2 errors, and that is the allowed limit." This is your last chance. MAKE THE MOST OF IT!

Go back to page 176.

Good Job!
"You have 1 Error"

I see you understand the previous concept! Let's take off and climb to 2000FT. We will fly directly to the international airport in our city. For this, we need to establish initial communication with the control tower. In this initial communication, there is a specific structure and content that we must adhere to. The order of the content is crucial. The structure should include the following:

Option 1: Aircraft registration, type of aircraft, origin, altitude, future intentions, and other relevant flight details. Continue on page 168.

Option 2: Type of aircraft and registration, flight intentions, altitude, and origin. Continue on page 163.

Option 3: Origin, registration, type of aircraft, endurance, flight intentions, and any other information the pilot considers relevant for air safety. Continue on page 139.

Good Job!

I see you understand the previous concept! Let's take off and climb to 2000FT. We will fly directly to the international airport in our city. For this, we need to establish initial communication with the control tower. In this initial communication, there is a specific structure and content that we must adhere to. The order of the content is crucial. The structure should include the following:

Option 1: Aircraft registration, type of aircraft, origin, altitude, future intentions, and other relevant flight details. Continue on page 160.

Option 2: Type of aircraft and registration, flight intentions, altitude, and origin. Continue on page 158.

Option 3: Origin, registration, type of aircraft, endurance, flight intentions, and any other information the pilot considers relevant for air safety. Continue on page 147.

STOP!

Incorrect response! This is your first error, but the exam has just begun. Remember that the maximum allowed is two errors, and at the third, we will stop, so you'll need to perform well in the remaining maneuvers to pass and obtain your license.

Don't worry, we will continue. We have landed on runway 04 after a successful 180° approach. We will now clear the runway, take the first taxiway, and return to the holding point for runway 04 for a new takeoff.

Remember that a rejected takeoff (RTO) is an emergency procedure and must be carried out from start to finish. It begins as soon as the decision is made, for any reason, to abort the takeoff. From there, you must reduce all power, maintain the runway axis, and bring the aircraft to a complete stop to then evaluate what happened.

As with any emergency situation, you will have priority on the runway, and other aircraft must wait until your operation is complete.

Let this be a lesson! Now, let's continue with your exam.

Proceed to page 149.

Good Job!
"You have 2 Errors"

I see you understand the previous concept! Let's take off and climb to 2000FT. We will fly directly to the international airport in our city. For this, we need to establish initial communication with the control tower. In this initial communication, there is a specific structure and content that we must adhere to. The order of the content is crucial. The structure should include the following:

Option 1: Aircraft registration, type of aircraft, origin, altitude, future intentions, and other relevant flight details. Continue on page 165.

Option 2: Type of aircraft and registration, flight intentions, altitude, and origin. Continue on page 117.

Option 3: Origin, registration, type of aircraft, endurance, flight intentions, and any other information the pilot considers relevant for air safety. Continue on page 131.

Great!
"You have 2 Errors"

The rejected takeoff (RTO) maneuver is extremely important for everyone's safety. After clearing runway 04, we will begin taxiing to a different runway and take off from runway 13.

Be cautious! This runway is in very poor condition, so we need to perform a short-field takeoff. Do you remember this maneuver? Let's review its procedure. We must extend the flaps. With the brakes applied, we set full power, release the brakes, and start the takeoff roll. My question is: this maneuver is performed with flaps extended in all cases, do you remember why?

Option 1: Extending the flaps increases lift and allows the aircraft to become airborne over a shorter distance during the takeoff roll. Continue on page 117.

Option 2: Extending the flaps delays the stall speed, so the aircraft can sustain at a lower speed and take off earlier, with a shorter takeoff roll than usual. Continue on page 156.

Option 3: Extending the flaps increases the wing area, which enhances lift and allows the aircraft to take off at a lower speed, resulting in a shorter runway distance required for takeoff. Continue on page 131.

STOP!

Incorrect answer! This is your first error, but the exam has just begun. Remember, the maximum allowed is two errors, and at the third, the exam will end. So you must focus on the remaining maneuvers to pass and obtain your license.

Remember that if you need to change position (enter a new position in this case), you must first analyze how many radials separate you from the new position. If the answer is less than 90 radials, you must perform a specific procedure. If the answer is exactly 90 radials, you must follow a different procedure. And if the answer is more than 90 radials, a different procedure is required. Review these theoretical concepts. In the Commercial Pilot exam, this will be VITAL!

Nonetheless, let this be a lesson! Now let's continue with your exam. Turn to page 177.

Great Job!
"You have 2 Errors"

You're doing very well so far. We've completed 18 maneuvers without errors, and we're almost done with your exam. Let's continue the flight. The control tower informs us that the QNH has changed to 1004 hPa and requests that we climb to 5000 feet. Be careful! The transition altitude is at 3000 feet. Tell me what steps to follow to comply with the air traffic controller's instructions...

Option 1: Begin the climb to 5000 feet. Upon passing 4000 feet (transition level), change the altimeter setting to 1013 hPa and continue climbing to reach 5000 feet. See page 156.

Option 2: Begin the climb to 5000 feet. After passing 3000 feet (transition altitude), change the altimeter setting to 1013 hPa and continue climbing to reach flight level 050. See page 176.

Option 3: Begin the climb to 5000 feet. After passing 3000 feet (transition altitude), change the altimeter setting to 1004 hPa and continue climbing to reach the desired altitude of 5000 feet. See page 117.

STOP!

Great, let's continue. The radial communication with the control tower is now established. Now, imagine that we are currently flying towards the airport on radial 150 and due to wind effect, the current heading is 300°. By direct instructions from air traffic control, we need to leave radial 150 and join radial 090. To do this, we must:

Option 1: Turn left to heading 360° and upon reaching radial 090, turn right to heading 270°, continuing the flight towards the airport. See page 154.

Option 2: Perform a turn either left or right to heading 360° and upon reaching radial 090, turn to heading 090°, continuing the flight towards the airport. See page 158.

Option 3: Perform a right turn to heading 360° and upon reaching radial 090, turn left to heading 270°, continuing the flight towards the airport. See page 171.

Well Done!

Excellent job! I believe there isn't much more to evaluate; you have the necessary knowledge to obtain your license. Let's perform one last maneuver to complete the exam. We are currently at flight level 050. Descend and execute a 360° approach to land on runway 13 at the airport. If you perform this maneuver correctly, we will have completed your exam!

So go ahead! Show me how to perform this maneuver step by step:

Option 1: The maneuver starts by overflying the beginning of runway 13 heading 310° at 1000 feet altitude. Then reduce power and turn to one side with a continuous descent until reaching the final approach for runway 13. See page 170.

Option 2: The maneuver starts by overflying the beginning of runway 13 at the altitude specified for the traffic circuit. Then reduce power and turn to one side with a continuous descent until reaching the final approach for runway 13. See page 169.

Option 3: The maneuver starts by overflying the beginning of runway 13 heading 130° at 1000 feet altitude. Then reduce power and turn to one side with a continuous descent until reaching the final approach for runway 13. See page 177.

STOP!

"You have 2 Errors"

Sorry! Your answer was incorrect! This is your second error and we haven't reached the sixth maneuver yet. You have reached the allowed limit. Radio procedures are a fundamental part of your career. Misinterpretation could cause incidents in congested airspace. You must master these techniques to become a professional pilot! Don't forget, they are a crucial part of your career!

Keep in mind that even small mistakes can lead to an accident, causing irreversible damage. We will continue from here with two errors, but don't be alarmed; if you perform all the remaining maneuvers correctly, your exam will be PASSED! So, stay encouraged and keep going!

"Remember that you have 2 errors, and that is the allowed limit." This is your last chance. MAKE THE MOST OF IT!

Go back to page 162.

STOP!
"You have 2 Errors"

Sorry! Your answer was incorrect! This is your second error and we haven't reached the sixth maneuver yet. You have reached the allowed limit. Statistics show that many accidents involving light aircraft are due to misinterpretation of instructions from air traffic control. Proper communication will help you reduce the margin of error, operating the flight more safely with the assistance of ATC.

Keep in mind that even small mistakes can lead to an accident, causing irreversible damage. We will continue from here with two errors, but don't be alarmed; if you perform all the remaining maneuvers correctly, your exam will be PASSED! So, stay encouraged and keep going!

"Remember that you have 2 errors, and that is the allowed limit." This is your last chance. MAKE THE MOST OF IT!

Go back to page 165.

Great Job!
"You have 1 Error"

Great job! I believe there's not much more to evaluate; you have the necessary knowledge to obtain your license. Let's perform one last maneuver to finish. We are at flight level 050. Let's descend and perform a 360° approach to land on runway 13 at the airport. If you can execute this maneuver correctly, we will complete your exam!

So, GO AHEAD! Show me how to perform this maneuver step by step:

Option 1: The maneuver starts by overflying the beginning of runway 13 on a heading of 310° at 1000 ft. Then reduce power and turn to one side with a continuous descent to the final of runway 13. Refer to page 168.

Option 2: The maneuver starts by overflying the beginning of runway 13 at the altitude specified by the traffic circuit. Then reduce power and turn to one side with a continuous descent to the final of runway 13. Refer to page 173.

Option 3: The maneuver starts by overflying the beginning of runway 13 on a heading of 130° at 1000 ft. Then reduce power and turn to one side with a continuous descent to the final of runway 13. Refer to page 175.

Excellent!

"You have 2 Errors"

Very well, let's continue. The radio communication with the control tower is established. Now, imagine that we are currently flying towards the airport on radial 150, and due to the wind effect, our current heading is 300°. By explicit instructions from air traffic control, we need to leave radial 150 and join radial 090. To do this, we should:

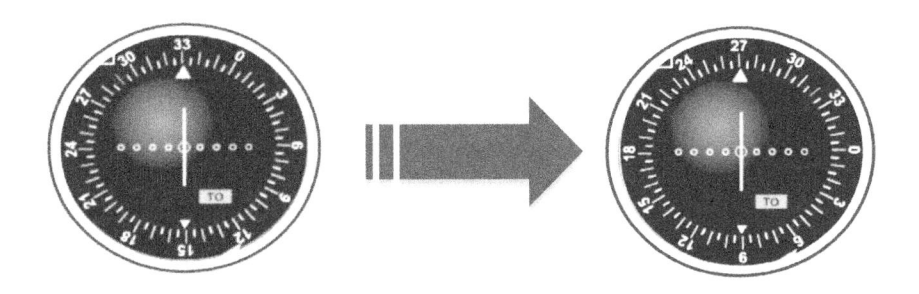

Option 1: Turn left to heading 360° and upon reaching radial 090, turn right to heading 270°, continuing the flight towards the airport. See page 157.

Option 2: Make a turn either left or right to heading 360° and upon reaching radial 090, turn to heading 090°, continuing the flight towards the airport. See page 117.

Option 3: Make a right turn to heading 360° and upon reaching radial 090, turn left to heading 270°, continuing the flight towards the airport. See page 159.

Great Job!

The RTO maneuver is of utmost importance for everyone's safety. After clearing runway 04, we will initiate another taxi to a different runway, taking off from runway 13.

Be careful! This runway is in very poor condition, so we need to perform a short field takeoff. Do you remember this maneuver? Let's review its execution. We need to extend the flaps. With the brakes applied, we apply full power, release the brakes, and begin the takeoff roll. My question is: this maneuver is performed with flaps extended in all cases. Do you remember why?

Option 1: Extending the flaps increases lift and allows the aircraft to take off over a shorter distance during the takeoff roll. See page 147.

Option 2: Extending the flaps delays the stall speed, allowing the aircraft to sustain at a lower speed and take off earlier with a shorter takeoff distance than usual. See page 148.

Option 3: Extending the flaps increases the wing area, which increases lift and allows the aircraft to take off at a lower speed, reducing the required runway distance. See page 154.

Excellent!

"You have 2 Errors"

We have landed on runway 04 after a successful 180° approach. We will clear the runway, take the first taxiway, and return to the holding point of runway 04 for a new takeoff. We enter runway 04, align the aircraft with heading 040°, and after applying full power, we begin the takeoff roll. However, during the roll, the airspeed indicator stops, and we decide to perform an RTO (rejected takeoff). To do this, we must:

Option 1: Reduce power, maintain the centerline of the runway, and bring the aircraft to a complete stop. Clear the runway and then evaluate the situation. See page 157.

Option 2: Decelerate the aircraft to a suitable speed for taxiing, return to the apron, and evaluate the situation. See page 117.

Option 3: With no remaining runway, continue with the takeoff normally, with the exception that we will enter the traffic pattern for an immediate landing in order to assess the cause of the malfunction. See page 134.

Excellent!
"You have 1 Error"

Great, let's continue. The radial communication with the control tower is now established. Currently, we are flying towards the airport inbound on radial 150, and due to the wind effect, our current heading is 300°. According to air traffic control instructions, we need to leave radial 150 and intercept radial 090. To do this, we must:

Option 1: Turn left to heading 360° and upon reaching radial 090, turn right to heading 270°, continuing the flight towards the airport. See page 149.

Option 2: Perform a turn either left or right to heading 360°, and upon reaching radial 090, turn to heading 090°, continuing the flight towards the airport. See page 162.

Option 3: Turn right to heading 360° and upon reaching radial 090, turn left to heading 270°, continuing the flight towards the airport. See page 177.

STOP!

Incorrect answer! This is your first error. Unfortunately, this was our last maneuver, and with this error, we need to lower your score.

Remember that the 360° approach involves performing a complete descent turn from 1000 feet down to landing, starting right over the runway threshold in use. Similar to the 180° approach, these maneuvers aim to practice gliding flight in abnormal situations, such as descending from 1000 feet to zero and performing a 360° turn.

Don't be alarmed. You've demonstrated great performance throughout the exam. Now, let's return to the aerodrome and land to complete your exam. Remember that study and practice lead to excellence in every pilot, and you must be aware that you've chosen a career that demands continuous improvement throughout your life.

Let's return and complete the task.

Refer to page 176.

Excellent!

Well done. We have landed on Runway 04 after a successful 180° approach. We will clear the runway, take the first taxiway, and return to the holding point for Runway 04 for a new takeoff. We enter Runway 04, align the aircraft with a heading of 040°, and after applying full power, we start the takeoff roll. However, during the roll, the airspeed indicator stops working and we decide to perform a rejected takeoff (RTO). To do this, we must:

Option 1: Reduce power, keep the aircraft on the centerline of the runway, and apply the brakes to come to a complete stop. Clear the runway and then assess what happened. Refer to page 166.

Option 2: Decelerate the aircraft to an appropriate taxiing speed, return to the ramp, and evaluate what happened. Refer to page 106.

Option 3: With no remaining runway, continue with the takeoff normally, with the exception of entering the traffic pattern for an immediate landing to assess the cause of the failure. Refer to page 155.

Great Job!

You're doing great so far. We've completed 18 maneuvers without errors, and we're almost finished with your exam. Let's continue the flight. The control tower informs us that the QNH has changed to 1004 hPa and requests that we climb to 5000 FT. Be careful! The transition altitude is at 3000 FT. Tell me the steps to follow to comply with the air traffic control instruction:

5000 FT

3000 FT

2000 FT

Option 1: Begin the climb to 5000 FT, and when passing 4000 FT (transition level), change the QNH to 1013 hPa and continue the climb to reach 5000 FT. Refer to page 154.

Option 2: Begin the climb to 5000 FT, and after passing 3000 FT, change the QNH to 1013 hPa and continue the climb to reach flight level 050. Refer to page 161.

Option 3: Begin the climb to 5000 FT, and after passing 3000 FT (transition altitude), change the QNH to 1004 hPa and continue the climb to reach the desired altitude of 5000 FT. Refer to page 151.

STOP!

Incorrect answer! This is your first error, but the exam has just begun. Remember, the maximum allowed is two errors, and the third will end the exam, so you need to be meticulous with the remaining maneuvers to pass and obtain your license.

Remember! Incorrect phraseology in your communications can lead to misunderstandings with air traffic control, which could trigger a series of events posing a risk to safety. We will emphasize your communications throughout this book, as they are crucial for everyone: for you as the pilot in command, for air traffic control, and even for other aircraft operating nearby. If you are unsure about a specific term or do not know how to communicate a particular action, seek assistance from air traffic control; they are there to ensure our flights are safe and comfortable.

Let this be a lesson! Now, let's continue with your exam.

Refer to page 168.

STOP!

Incorrect answer again! This is your second error. Unfortunately, since this was our last maneuver, we must lower the overall score.

Remember that a 360° approach involves a complete descent from 1000 FT to landing, starting directly over the runway threshold. Like the 180° approach, these maneuvers are designed to practice gliding flight under abnormal conditions, such as descending from 1000 FT to zero while executing a 360° turn.

Don't be discouraged. You performed well throughout the exam. Let's return to the aerodrome and land to complete your exam. Remember, study and practice are key to excellence in piloting, and you must be aware that you've chosen a career that requires continuous improvement throughout your life.

Let's return and finish the task.

Refer to page 178.

Excellent!
"You have 1 Error"

You're doing very well so far. We've completed 18 maneuvers without errors, and we're almost at the end of your exam. Let's continue the flight. The control tower has informed us that the QNH has changed to 1004 hPa and requests that we climb to 5000 FT. Be careful! The transition altitude is 3000 FT. Tell me what steps to follow to comply with the air traffic control instructions...

5000 FT

3000 FT

2000 FT

Option 1: Begin climbing to 5000 FT. Upon passing 4000 FT (transition level), change the QNH to 1013 hPa and continue climbing to reach 5000 FT. Refer to page 152.

Option 2: Begin climbing to 5000 FT. After passing 3000 FT, change the QNH to 1013 hPa and continue climbing to reach flight level 050. Refer to page 164.

Option 3: Begin climbing to 5000 FT. After passing 3000 FT (transition altitude), change the QNH to 1004 hPa and continue climbing to reach the desired altitude of 5000 FT. Refer to page 168.

Excellent!
"You have 2 Errors"

You'Great job! There isn't much more to evaluate; you have the necessary knowledge to obtain your license. Let's perform one final maneuver to conclude. We will continue at flight level 050, descend, and execute a 360° approach to land on runway 13 at the airport. If you complete this maneuver correctly, we will have finished your exam!

So, GO AHEAD! Show me how to perform this maneuver step by step:

Option 1: The maneuver begins by overflying the beginning of runway 13 on a heading of 310° at 1000 FT altitude. Then, reduce power and turn in one direction with a continuous descent until reaching the end of runway 13. Refer to page 117.

Option 2: The maneuver begins by overflying the beginning of runway 13 at the altitude described for the traffic pattern. Then, reduce power and turn in one direction with a continuous descent until reaching the end of runway 13. Refer to page 178.

Option 3: The maneuver begins by overflying the beginning of runway 13 on a heading of 130° at 1000 FT altitude. Then, reduce power and turn in one direction with a continuous descent until reaching the end of runway 13. Refer to page 156.

Congratulations!

Your exam has successfully concluded. You made it to the end with only one error, which is an excellent average. There's no doubt that completing an exam requires a certain level of pressure management, and you've mastered it. It will be a true honor to share the cockpit with you if aviation brings us together again. Here is your Private Pilot's License diploma. I know you will carry it with pride and respect for our beautiful profession. Remember, continuous study makes every pilot better each day!

Capt. Thómas

CERTIFICATE
OF APPROVAL

The Aeronautical Academy certifies that you have successfully completed your training in the PRIVATE PILOT LICENSE (PPL) course.

Having passed all the required exams for each module, this certificate is issued as sufficient proof and endorsement to be presented to the relevant authorities, certifying that you have acquired the knowledge required by the course.

Dirección Académica
Biblioteca Aeronáutica

PILOTO PRIVADO
DE AVIÓN

Congratulations!

Your exam has concluded with great success. You have reached the end without making any mistakes, which fully deserves my recognition of your effort and capability as Pilot in Command. It will be a true honor to share the cockpit with you if aviation brings us together again. Here is your Private Pilot's License diploma. I know you will carry it with pride and respect for our beautiful profession. Remember, continuous study makes every pilot better each day!

Capt. Thómas

CERTIFICATE
OF APPROVAL

The Aeronautical Academy certifies that you have successfully completed your training in the PRIVATE PILOT LICENSE (PPL) course.

Having passed all the required exams for each module, this certificate is issued as sufficient proof and endorsement to be presented to the relevant authorities, certifying that you have acquired the knowledge required by the course.

Dirección Académica
Biblioteca Aeronáutica

PILOTO PRIVADO
DE AVIÓN

Colección HOW DOES IT WORK?

Baeronautica.com

Congratulations!

Your exam has concluded successfully. You have reached the end with only two errors, which is a good average. Without a doubt, handling an exam requires managing a certain level of pressure, and you have overcome it. It will be a true honor to share the cockpit with you if aviation brings us together again. Here is your Private Pilot's License diploma. I know you will carry it with pride and respect for our beautiful profession. Remember, continuous study makes every pilot better each day!

Capt. Thómas

CERTIFICATE
OF APPROVAL

The Aeronautical Academy certifies that you have successfully completed your training in the PRIVATE PILOT LICENSE (PPL) course.

Having passed all the required exams for each module, this certificate is issued as sufficient proof and endorsement to be presented to the relevant authorities, certifying that you have acquired the knowledge required by the course.

Dirección Académica
Biblioteca Aeronáutica

Colección HOW DOES IT WORK?
PILOTO PRIVADO
DE AVIÓN

Made in United States
Orlando, FL
07 May 2025